T0330889

The Electronics Industry Research Series

THE Japanese
ELECTRONICS
INDUSTRY

Wataru Nakayama
William Boulton
Michael Pecht

CRC Press
Taylor & Francis Group
Boca Raton London New York

CRC Press is an imprint of the
Taylor & Francis Group, an **informa** business

Library of Congress Cataloging-in-Publication Data

Nakayama, Wataru.
    The Japanese electronics industry / Wataru Nakayama, William
Boulton, Michael Pecht.
        p.   cm.
    Includes bibliographical references and index.
    ISBN 1-58488-026-0 (alk. paper)
    1. Electronic industries--Technological innovations--Japan.
I. Boulton, William. II. Pecht, Michael. III. Title.
HD9696.A3J36754 1999
338.4'762138l'0952—dc21   99-28223
                                                            CIP

PREFACE

The growth of the Japanese electronics industry in the increasingly integrated world economy has been driven by a combination of market forces and the unique characteristics of the Japanese social organization and people. The Japanese electronics industry, as an industrial phenomenon, has received considerable attention from researchers in various fields: economists, technology watchers, historians, market researchers, journalists, and government-sponsored study teams. Broadly categorized in the light of their foci, the existing literature about the Japanese electronics industry falls into two groups. The studies in one category are the analyses of the historical development of the electronics industry in Japan. The purpose of such studies is to find out the secret of the enormous commercial success achieved by the industry. The other type consists mainly of reports published in the United States, which focus on the issue of America's competitiveness in the face of challenges from Japanese technology.

The electronics industry is now a leading component of the Japanese industrial infrastructure, and thus an indispensable topic in the study of Japanese industry and companies. Moreover, the electronics industry is rooted in Japanese society, so that it is often discussed in studies of Japanese society and culture. When the information dispersed in published books and papers about Japan, her industries and her electronics industry are combined, the available knowledge about the subject is voluminous. Why another book on the Japanese electronics industry?

For one reason, the rapid pace of technological development and the concomitant change in the industrial landscape require frequent updating of the relevant information. For another, the study of such a complex subject demands multiple viewpoints because it has to deal with technical, social, and cultural matters. Because of the social and cultural implications, no study can be free from bias implanted in investigators cultivated by their own upbringing and current environment. Most of these books and study reports available to readers in the West were written by experts who stayed in or visited Japan, interviewed key players of the industry, and studied government and corporate statistics. Those who became acquainted with corporate gurus tend to praise the achievements of the Japanese electronics industry. If the investigator is sent from a foreign company in danger of extinction in this highly competitive market, his or her report could be critical of Japanese practices.

The first author of the present book is Japanese and worked for almost twenty years in a Japanese company, taught at a Japanese university for more than six years, and came to work for an American university. With such personal experience, our hope is to portray the Japanese electronics industry from the angle of an insider but with a conscious attempt to be an objective observer, as difficult as that may be. The second author, Dr. William Boulton, has lived and worked in Japan as an executive, professor, and research scholar, and has authored numerous articles and reports on Japan and its electronics industry. The third author, Dr. Michael Pecht, has been leading research efforts on Asian electronics and published several books on the subject.

This book is intended for general readers. Chapters 1 and 2 provide background information. Chapter 3 describes the history of the Japanese electronics industry and attempts to illuminate its characteristics. Chapter 4 presents the current state of the industry based on a collection of data. Chapter 5 deals with research, development and education that are important for the future of the industry.

Acknowledgments

Professor Nakayama deeply thanks his friends in Japan for providing him with useful information and materials. Mr. Haruo Kozono, General Manager, Computer Systems Integration, Sony Corporation, sent the author various materials about the history of the Japanese electronics industry, SONY, and recent technical developments in Japan. Dr. Masaru Ishizuka, Chief Research Scientist, Energy and Mechanical Research Laboratories, Toshiba Corporation, provided the authors with publications on Toshiba's mini-notebook computer 'Libretto'. Dr. Kazumasa Fujioka, Senior Researcher, Mechanical Engineering Research Laboratory, Hitachi, Ltd., sent the information regarding the current state of the Japanese electronics industry. Without their cooperation this book could not have been completed. Neither would it have been possible for Professor Nakayama to co-author this book without the support and encouragement from his wife Michiko. Although the author takes for granted the customary Japanese wife-husband relationship, he feels deeply indebted for it from the bottom of his heart. Finally, the authors would like to thank U.V. Ramgopal and Meerra Ganeshan for organizing and formatting the book.

The Authors' Profiles

Wataru Nakayama received his Doctor of Engineering degree from the Tokyo Institute of Technology in 1966. After a three-year stint at Canadian universities, he returned to Japan in 1970 to work for Hitachi, Ltd. He worked at Hitachi's Mechanical Engineering Research Laboratory before moving to the Tokyo Institute of Technology in 1989. His primary function at Hitachi was the conduct and supervision of heat transfer research for various products. Later he broadened his area of interest to interdisciplinary subjects, and was promoted to the rank of Honorary Engineer at the end of his association with the company. At TIT he taught and conducted research on microelectronic packaging, with a focus on power and thermal management of computers, which required multidisciplinary expertise. He retired from TIT in March 1996. Since June 1996, he has been associated with the CALCE Electronic Packaging Research Center, the University of Maryland, as a visiting professor.

Dr. Nakayama has been active in international professional societies. He has authored and co-authored some 200 papers. He received prominent awards, including the Heat Transfer Memorial Award from ASME in 1992 and best paper awards from ASME, JSME, and other societies. He served as Chairman of the Heat Transfer Society of Japan in 1994, Chairman of ASME Japan in 1990/92, and Chairman of JSME Thermal Engineering Division in 1990, and is currently serving as a member of the Executive Committee of the International Center for Heat and Mass Transfer. He co-organized a number of international conferences and delivered keynote speeches and seminars at conferences and universities. He is an ASME Fellow and IEEE Senior Member.

William R. Boulton received a doctorate in Business Strategy and Policy for the Harvard Business School in 1977. He received undergraduate and masters degrees in Business Administration from the University of Washington and then worked for General Telephone International in Hong Kong and Japan. After completing his doctorate, Dr. Boulton taught at the University of Georgia, where he became the director of graduate programs for the College of Business Administration. In 1986, he became a Fulbright research scholar in Japan where he conducted research on advanced manufacturing and robotics technologies. In 1990, he moved to Auburn University as the C.G. Mills Professor of Strategic Management and the director of the Center for International Commerce. Since 1992, he has been a member of three research teams and edited three reports on electronics manufacturing technologies and industrial development policies of Japan and other Asian countries.

Michael Pecht is the Director of the CALCE Electronic Products and Systems Consortium (EPSC) at the University of Maryland and a Full Professor with a three way joint appointment in Systems Engineering, Mechanical Engineering, and Engineering Research. Dr. Pecht has a BS in Acoustics, a MS in Electrical Engineering and a MS and PhD in Engineering Mechanics from the University of Wisconsin. He is a Professional Engineer, an *IEEE* Fellow, an ASME Fellow and a Westinghouse Fellow. He has written eleven books on electronics products development. He served as chief editor of the *IEEE* Transactions on Reliability for eight years and on the advisory board of *IEEE* Spectrum. He is currently the chief editor for *Microelectronics Reliability International.* He serves on the board of advisors for various companies and provides expertise in strategic planning in the area of electronics products and systems development.

TABLE OF CONTENTS

Preface

Acknowledgements

Chapter 1

Japan Overview

1.1. Land

An arc of islands lying east of the Asian continent forms Japan, stretching for about 2,600 km from northeast to southwest. Japan is slightly smaller than the state of California, with 145,868.72 sq. miles (377,800 sq. kilometers) of land area. The four major islands include Hokkaido to the north, the central island of Honshu, and Kyushu and Shikoku to the south.

Surrounded by seas, Japan has an unusually long coastline in comparison with its area. The area contains volcanic and earthquake belts spreading across the land and its coastal waters. The activities of these belts have combined to create the scenic beauty of a country with over twenty-three national parks, and over eleven hundred hot springs. Since the rivers are short and rapid, they are seldom suitable for transportation. About 70 percent of its land is covered with forests, predominantly with trees in the north, deciduous, broad-leaved trees in the central part, and broad-leaved, evergreen trees in the south. Only one-fifth of the entire land area of the country is flat, characterized by plains, basins, and tablelands. The fertile plains, generally used to raise rice, are located at the lower ends of the large rivers.

Agriculture in Japan primarily involves growing rice, which is not only the nation's staple food but also was once the basic unit of economic value. The size of a feudal fief was expressed in terms of how many koku - a measure of grain roughly equal to 180 liters—of rice the land produced. Rice in Japan had an economic function that approximated the role that gold played in Europe. Japan's agricultural history is about rice and the technological innovations to increase rice production. Land reforms after World War II took away land from non-farming landowners and gave it to the many tenant farmers who actually tilled it. The new independent farmers improved the productivity of rice, which was in short supply, and improved Japan's agricultural self-sufficiency.

In 1997, small-scale farming characterized Japan's high-cost agriculture and limited its productivity improvement. Four years of bumper crops led to the largest rice surplus since 1980 and a decrease in rice prices. Japan's overall self-sufficiency in calorific supply was only 42 percent in 1996, down from 79 percent in 1960. It had reached the lower end of the 41 percent to 46 percent self-sufficiency target set by the Ministry of Agriculture, Forestry, and Fisheries for 2005. Japan's self-sufficiency for commodity foods is 63 percent. Only vegetables, eggs, milk, and other dairy products are above the average for self-sufficiency. This situation has increased the need to import food to meet the diversified diet and eating habits of the Japanese.

1.2. Climate

Japan has four distinct seasons. Since the archipelago stretches over 1,800 miles from north to south, the climate varies greatly. The northern end of Japan has the same latitude as Quebec, Canada, while the southern end has the same latitude as Key West, Florida. The temperature differences between Hokkaido in the north and Okinawa in the south can range from zero to 30 degrees C. Siberian winds in the winter and hot Pacific winds in the summer create extreme differences in climate between the Japan Sea coast and the Pacific. The Japan Sea coast is dry in the summer and has heavy, moist snowfall in the winter. The Pacific coast is hot and humid in the summer and cold and dry in the winter. Most of Japan also has a rainy season from mid-June to mid-July, and frequent typhoons from August through October.

1.3. Population and Ethnic Groups

Japan's population as of February 1996 was 125.7 million. The largest Japanese cities are Tokyo (7.961 million), Yokohama (3.317million), Osaka (2.601 million), Nagoya (2.151 million), Sapporo (1.771 million), Kyoto (1.464 million), Kobe (1.42 million), Fukuoka (1.293 million), Kawasaki (1.208 million), Hiroshima (1.113 million), and Kitakyushu (1.18 million) as of July 1996. The capital, Tokyo, has a population during the working day of about 15 million.

1.4. Language

Japanese is the national language, and the only one spoken universally throughout the country. English is taught as a second language in schools, but with limited success. Relatively few Japanese are comfortable speaking English, though it is often found in train stations.

Written Japanese uses a combination of syllabic symbols (kana) and characters borrowed from the Chinese (kanji), so it is impossible for Westerners to read Japanese without training. Numerous books and software programs provide lessons in basic Japanese. In spoken Japanese, many Western words have been borrowed and are spoken in "Japlish"; an often shortened version with Japanese pronunciation.

1.5. Religion

The two major religions in Japan are Buddhism and Shinto. Shinto, "the way of the gods," makes no great demands on the worshipper, who makes offerings to ancestors. It involves the purification of the worshipper's hands and mouth, pulling on the shrine's hemp rope and clapping hands to attract the divinity's attention, prayer, giving of money, and purchasing an amulet. Zen Buddhism is practiced privately through the practice of meditation, derived from India and China. Japan owes to Shinto the firm involvement of the individual in the family and the state, given dogmatic form and a political dimension by Confucianism. Buddhism has added the spiritual and speculative element, with a touch of transcendence and a dawning of personal awareness.

1.6. Education

Japan, with a literacy rate of over 98 percent, has one of the best-educated work forces in the world. Education from elementary through upper secondary school is conducted in accordance with teaching guidelines determined by the government; textbooks compiled by private firms in conformity with these guidelines must obtain the authorization of the Ministry of Education. Each locality chooses its textbooks from those that have received the ministry's approval. The rigid education system, based on memorization, is blamed for the lack of innovation and creativity needed for rapidly changing societies. Japan's lack of Nobel Prize winners is typically referenced to substantiate this claim. Through 1990, Japan had received only 5 Nobel Prizes, compared to 128 for the U.S.A., 40 for the U.K., and 21 for Germany. A report issued in 1987 by the Provisional Council on Educational Reform emphasized the importance of respecting students' individuality, preparing students for life in an increasingly international and information-oriented society, and promoting continuing education. Curriculum revisions have been introduced since 1990 to upgrade the teaching of foreign cultures, history, and languages to prepare students for internationalization.

Japanese education is based on a six-three-three-four year system. Compulsory education, extending from age six to fifteen, is completed by 99.98 percent of school-aged children. Families must then pay tuition for high school education. As of 1997, 96.8 percent of Japanese children advanced to high school. While the population of 18 year-olds continues to decline, the percentage of those seeking higher education has been rising. Reforms in the education system now allow "grade-skipping" for high school juniors to go on to college, and the reduction of the school week to five days.

The percentage of Japanese students advancing directly into higher education (including universities, colleges and junior colleges or vocational schools) reached 40.7 percent in 1997 as compared with 31 percent in 1987. Exams require many students to delay entry for one or more years after high school in order to pass at levels acceptable for admission into a specific university.   In 1997, 47.3 percent of students eventually entered undergraduate university programs or regular courses at junior colleges, up from 36.1 percent in 1987. As of the spring of 1997, 26 percent of female students entered the university and 22.9 percent entered junior colleges, surpassing the number of males entering advanced programs by over three percentage points.

1.7. Government Structure

Japan's government is a constitutional monarchy. The Emperor (His Majesty Emperor Akihito) is the symbol of state; the chief executive is the Prime Minister (Keizo Obuchi in 1998). The present system is built upon the separation of powers of three branches of government (legislative, executive, and judicial), which act to check and balance one another.

Japan adopted a parliamentary democratic system, in which the Diet, the sole legislative body of the state, is the highest organ of government power. The Diet consists of two houses, the House of Representatives (500 seats, 4 year term) and the House of Councilors (252 seats, 6 year term).

Japan has a parliamentary cabinet system, in which executive power is vested in the cabinet. The cabinet consists of the prime minister and twenty ministers of state and is collectively responsible to the Diet. The prime minister, who in practice is always a member of the House of Representatives, is elected by the members of the Diet. The National Government Organization Law stipulates that the administration shall consist of the Prime Minister's Office and twelve ministries, including Foreign Affairs; Justice; Finance; Education; Health and Welfare; Agriculture, Forestry and Fisheries; International Trade and Industry; Transport; Posts and Telecommunications; Labor; Construction; and Home Affairs. In addition to these ministries, there are agencies, including the National Public Safety Commission; the Management and Coordination; Hokkaido Development; Okinawa Development; Defense; Economic Planning; Science and Technology, Environment; and National Land, and other administrative bodies, including the National Personnel Authority, and the Board of Audit. As of the end of March 1995, there were approximately 1,164,000 national public employees, including the Self-Defense Force.

There are five categories of courts in Japan: the Supreme Court, high courts, district courts, family courts, and summary courts. In addition to these, there is a court of impeachment in the Diet to try judges.

1.8. Modern History

After nearly eighty years of contact with Europeans, chiefly the Portuguese, Spanish, and Dutch, the Tokugawas (1603-1867) resolved to have no more to do with outsiders. Japan embarked on a period of isolation and focused on building its national social structure, which provided the infrastructure for its development in the second half of the nineteenth century. Japan's isolation was finally broken by the arrival of Commodore Perry, an American, in 1853. The Tokugawa Period ended in 1868 with the restoration of the emperor. The opening of Japan led to the modernization of its administrative and economic structure and its quest for world power. The conquests of Korea, Manchuria, China, and Southeast Asia were relinquished with Japan's defeat in World War II.

Today, Japan has become the dominant economic power in Asia and is a key part of the U.S. security structure in Asia. Japan began facing severe financial setbacks with the end of the "bubble" economy in 1989, compounded by the Kobe earthquake in 1995 and the near financial collapse of South Korea, Thailand, and Indonesia in 1998. Nearly 40 percent of the debt in Asia is held by Japanese banks, which has caused serious restructuring of Japan's financial and banking system. On October 12, 1998, Japanese parliament passed new legislation allowing government to nationalize banks that have failed. The Long Term Credit Bank of Japan was first to be taken over as Japan's Resolution Trust Corporation assumed its bad debts. On December 12, Nippon Credit Bank was placed under state control. A new Financial Revitalization Committee oversees the restructuring. Nearly $2 trillion was set aside to rescue failing banks, and $3 trillion to rescue deposition of failed banks.

Chapter 2

Economic Overview

After World War II, Japanese industrial policies were directed at rebuilding its manufacturing base. By 1960, Japan completed its reconstruction of basic industries and began to aggressively export products into world markets. Shortages of raw materials caused Japan, as a nation, to develop an export orientation that would generate the foreign currency needed to pay for imports of raw materials. In planning for future export opportunities, Japan's industrial policy shifted from reconstruction of mature industries to developing new, high-technology growth industries that could provide growth opportunities for the future. Through the 1960s and 1970s, a strategy was implemented for establishing Japan's technology base. The strategy broadened during the 1980s to make Japan the world's technological leader. The "Japan, Inc." that had grown out in the 1950s and 1960s was being directed into "Japan Tech," a future leader in knowledge-based industries. However, with the end of the Cold War and the collapse of the Japanese "Bubble Economy" in 1989, competition from other Asian countries exploded, while Japanese companies consolidated operations and cut back on expenses that included technology costs.

The currency crisis in Southeast Asia raised serious questions about the government-led type of economy that directed Japan's early development success. Industrial policy changes required for a more developed economy in the face of greater regional and global competition have been slow to come. In the early 1980s, the U.S. criticized Japan's industrial policy (1) for targeting specific industries and technologies and providing low-interest loans, technology subsidies, and tax privileges; (2) for curbing imports through administrative guidance; and (3) for supporting cartels. Japan's continued trade surplus suggested that free and open competition was restricted, while Japanese industry also relocated less competitive manufacturing facilities to other countries.

2.1. Philosophy of Economic Policy

The economic policy of Japan is moving towards deregulation, since over-regulation is considered one of the many key structural problems confronting Japanese social systems. The move towards market-oriented activities is considered essential in a world of free competition. Japan's involvement in directing industry and providing safety nets and protection is slowly being phased out. Since Japan's economy is one of the largest in the world, deregulation is essential to revitalise private-sector economic activities. Japan's success with regulation has slowed efforts to stimulate creativity and innovation in business. The government's Deregulation Committee has targeted some eleven thousand regulations being targeted for elimination

The bursting of the Bubble Economy in 1989 has led to serious problems in Japan's financial and economic systems. For the 1997 fiscal year ending March 31, 1998, Japan's nineteen top "city" banks wrote off a total of 10 trillion Yen (nearly $70 billion) in problem loans, as compared with 6.17 trillion Yen in bad loans for 1996. The Financial System Reform Bill, taking effect on December 1, 1998, revises twenty-two financial laws and two tax laws, liberalizes brokerage commissions by the end of 1999, and lifts restrictions on securities subsidiaries of banks by mid-1999. Licensing for broker-dealers will be abolished in favor of a registration system. Additional approvals will be needed for underwriters, derivatives, and proprietary trading companies. Restrictions against vertical integration between banks, security firms, and insurance firms will be phased out in 2001.

2.2. Current Economic Conditions

The slowdown in Japan's Gross Domestic Expenditures resulted in the great economic recession of 1998. The national budget deficit for fiscal 1997 (ending March 1998) was $10.8 billion. On July 12, 1998, Japanese voters responded by voting against the ruling Liberal Democratic Party (LDP). Small businessmen, farmers, merchants, office workers, and others displayed their disappointment with the government's lack of action, causing Prime Minister Ryutaro Hashimoto to resign. There was little optimism, however, that a successor would do a better job. The consensus-driven political process had been unable to arrive at fruitful actions for over five years. The international community was concerned that Japan's inability to respond would lead to the collapse of its economy, taking down the already shaky Southeast Asian economies, forcing devaluation of China's currency, causing a downturn in the U.S. and European economies and leading to a global business slump.

The economic problems of Japan had been building since 1989. Gross domestic expenditures (GDE) grew less than 10 percent between 1992 and 1997 in both nominal and real terms; private expenditures grew less than 5 percent. Much of the blame for such poor economic performance was laid on the government's economic policies. After adding a national sales tax in the early 1990s, it increased the national sales tax from 3 percent to 5 percent in April 1997. It then increased the percentage of medical costs that salaried workers had to pay from 10 percent to 20 percent in September 1997. *Asiaweek* reported that the nominal GDE actually declined 0.2 percent for the twelve month period ending in June 1998 [*Bottom Line* 1998]. In a survey taken

Table 2-1 Japan's Gross Domestic Expenditures

Year	Nominal GDE (billion Yen)	Nominal GDE Growth (%)	Real GDE (billion Yen)	Real GDE Growth (%)	Private Expenditures (billion Yen)
1992	471,064	2.8	450,924	1.0	57.8
1993	457,381	0.9	452,282	0.3	58.6
1994	479,260	0.8	455,197	0.6	59.7
1995	483,220	0.8	461,894	1.5	60.2
1996	499.861	3.4	480,013	3.9	59.8
1997	507,271	1.5	484,318	0.9	60.6

Source: *National Income*, NIPPON 1998: **Business Facts and Figures**, Japan External Trade Organization, June 1998, p. 16.

between April and June 1998, the *Far Eastern Economic Review* found that over 80 percent of the respondents were pessimistic about Japan's economic future–more pessimism than in Indonesia (70 percent), South Korea (65 percent), or Thailand (55 percent). Fewer than 15 percent of the respondents felt that their political leaders made decisions with the public interest in mind. *The Far Eastern Economic Review* projected a growth rate of 0.3 percent for 1998. Economic recovery was not expected until after the year 2000 [*Economic Indicators* 1998]. Table 2-1 gives Japan's gross domestic expenditures.

2.3. Foreign Trade

Global trade showed a major slowdown, falling from 20.4 percent growth in 1995 to just 3.3 percent in 1996. The slowdown in trade occurred at a time when the world's GDP increased its growth rate from 3.7 percent in 1995 to 4.0 percent in 1996. Despite the drop in the value of Yen, Japan's export value dropped 0.7 percent in 1996, compared with 1995. The unit value of exports (in Yen), which had fallen throughout the early 1990s, rose by 8.5 percent in 1996 as producers raised prices with the Yen's depreciation. The volume of imports, reflecting the depreciation in value of the currency, rose only 2.3 percent in 1996, compared with 12.5 percent in 1995. The unit value of imports also increased 17.7 percent. Figure 2-1 shows the primary trade pattern between Japan and its trading partners in Asia, the US, and the EU.

As shown in Table 2-2, Japan maintained a current accounts surplus during its economic recession, though it declined between 1994 and 1996. Rising exports improved the current account balance for 1997. Major export categories for 1997 included general machinery (12.1 trillion Yen), electrical machinery (12.0 trillion Yen), and transportation equipment (11 trillion Yen). Automobiles were the leading export (7.1 trillion Yen), with primary destinations in the U.S., Australia, and Germany. Over 3.8 trillion Yen in office equipment, such as copiers, was exported to countries like the U.S., Germany, and the Netherlands. Scientific and optical equipment, including cameras, shipped to countries like the U.S., Germany, and Taiwan, amounted to over 2.2 trillion Yen. Heavy industry, including ships, heavy equipment, and steel, accounted for another 4.6 trillion Yen in exports. The U.S. is the main destination for all these exports.

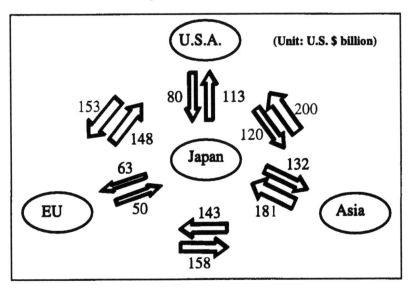

Figure 2-1 Value of trade between Major Regional Economical Blocs (1996)
Source: NIPPON 1998: Business facts & figures, JETRO, June 8, 1998, p. 65.

Table 2-2 Japan's International Balance of Payments

(billion Yen)	1994	1995	1996	1997
Current balance	13,342.5	10,386.2	7,157.9	11,435.7
Balance of trade	14,732.2	12,344.5	9,096.6	12,329.1
Exports (% change)	39,348.5 (0.5%)	40,259.6 (2.3%)	43,565.9 (8.2%)	49,517.2 (13.7%)
Imports (% change)	24,616.6 (3.9%)	27,915.3 (13.4%)	34,469.3 (23.5%)	37,188.0 (7.9%)
Balance of services	-4,897.6	-5,389.8	-6,779.2	-6,526.0

Source: NIPPON 1998: Business facts & figures, JETRO, June 8, 1998, p. 60.

2.3.1. Japanese Exports

An increase in offshore production by Japanese companies is shifting Japanese production base to high-value added products, which have lower price elasticity and a smaller volume market demand. Japan's high-quality, high -value export strategy is expected to sustain its export position throughout the 1990s, even though growth may be more limited. The top ten export areas from Japan are shown in Table 2-3.

Automobiles and related parts have been the primary areas of export, followed by electronics-related products, such as semiconductors and business machines. High-precision optical and imaging equipment have

Table 2-3 Top Ten Export Commodities (1993-1996) (hundred million Yen)

Rank	1993	1994	1995	1996
	Automobiles 65,505	Automobiles 58,366	Automobiles 49,797	Automobiles 55,138
2nd	Business Machines 32,703	Semiconductors and Other Electronic Components 29,957	Semiconductors and Other Electronic Components 38,299	Semiconductors and Other Electronic Components 38,812
3rd	Semiconductors and Other Electronic Components 24,458	Business Machines 29,790	Business Machines 28,892	Business Machines 31,885
4th	Motor Vehicle Parts 16,544	Motor Vehicle Parts 17,982	Motor Vehicle Parts 17,815	Imaging Equipment 18,945
5th	Steel 16,139	Optical Instruments 16,266	Optical Instruments 17,358	Motor Vehicle Parts 18,405
6th	Optical Instruments 15,849	Steel 15,199	Steel 16,443	Steel 16,548
7th	Primary Motors 12,730	Primary Motors 14,346	Primary Motors 14,503	Primary Motors 15,333
8th	Imaging Equipment 11,822	Ships 11,290	Organic Compound 10,317	Organic Compound 11,057
9th	Ships 11,348	Imaging Equipment 11,016	Ships 9,981	Ships 9,953
10th	Communications Equipment 9,035	Communications Equipment 8,793	Imaging Equipment 9,632	Imaging Equipment 9,727

Source: Kanzei Nempo (Annual Report of Customs and Tariff), Customs and Tariff Bureau, Ministry of Finance (June 30, 1997)

Table 2-4 Japan's Exports by Region (1991-1996) (million Yen)

Year	Total Exports	Asia	Europe	North America	South America	Africa	Oceania
1991	42,359,893	15,679,811	9,719,724	14,481,386	540,733	812,764	1,125,455
1992	43,012,281	16,604,209	9,404,477	14,368,975	654,102	857,751	1,122,767
1993	40,202,449	16,412,063	7,544,846	13,691,466	632,972	831,796	1,089,246
1994	40,497,553	17,170,506	6,932,115	13,950,405	613,787	715,951	1,114,748
1995	41,530,895	18,910,714	7,214,247	13,107,292	616,320	703,761	978,545
1996	44,731,311	20,755,620	7,565,499	14,081,238	603,899	637,980	1,087,075

Source: *Kanzei Nempo* (Annual Report of Customs and Tariff), Customs and Tariff Bureau, Ministry of Finance (June 30, 1997).

Table 2-5 Growth in Exports of Electrical Equipment as Capital Goods (Top 10 Products)

	1990 (% Share)		1996 (% Share)		Growth %	Contribution %
1. ICs and microassemblies	1,223.6	(15.1)	2,690.2	(28.2)	119.9	18.1
2. Semiconductor devices & photosensitive devices.	293.6	(3.6)	566.3	(5.9)	92.9	3.4
3. Thermionic tubes, cold cathode tubes, photocathode tubes.	417.5	(5.2)	624.7	(6.5)	49.6	2.6
4. Electric accumulators	109.3	(1.3)	241.4	(2.5)	120.9	1.6
5. Electrical capacitors	180.5	(2.2)	300.9	(3.2)	66.8	1.5
6. Electric apparatus for switching or protecting circuits under 1000V	335.8	(4.1)	453.6	(4.7)	35.1	1.5
7. Electrical circuit components	78.5	(1.0)	155.1	(1.6)	97.6	0.9
8. Printed circuits	72.4	(0.9)	149.0	(1.6)	105.6	0.9
9. Oscilloscopes & measuring devices	95.1	(1.2)	162.5	(1.7)	70.7	0.8
10. Electrical control boards	156.2	(1.9)	202.8	(2.1)	29.8	0.6
Subtotal of top 10	2,962.5	(36.6)	5,546.5	(58.1)	87.2	31.9
Total of electrical equipment	8,097.4	(100.0)	9,550.7	(100.0)	17.9	17.9

Note: Growth and Contribution are comparisons of two time periods, 1990 and 1996.

Source: JETRO White Paper on International Trade 1997: Global trade in the era of information communications (Japan External Trade Organization, 1997) p. 36.

been major export areas. Japan continues voluminous exports in heavy industries like steel, primary motors, and ships.

The high cost of Japanese manufacturing, combined with the relative slowdown in Japan's rate of technology development has allowed other Asian countries to close the competitive gap with Japan. The transfer of electronic manufacturing from Japan to other countries stimulated demand for Japanese electronic components and assembly equipment. The rapid expansion of computer-related manufacturing in Taiwan, Singapore, Malaysia, China, and the Philippines increases the market for Japan's components and electronic manufacturing equipment as shown in Table 2-4; exports to Asia increased from 15.7 billion Yen to 20.8 billion Yen between 1991 and 1996. Exports to other regions have been relatively stable.

Table 2-5 shows exports playing the lead role in electronics. These included integrated circuits and microassemblies, semiconductor devices and photosensitive semiconductor devices, thermionic tubes, cold cathode tubes, and photocathode tubes. In addition, offshore investment by Japanese firms has increased exports of capital equipment and electronic components as production parts for these subsidiaries [JETRO White Paper on International Trade 1997].

The value of overseas plant contracts made by Japanese companies in 1997 fell 40.8 percent to $11.7 billion, the lowest figure since 1991. The number of contracts dropped to 924, compared to 1129 in 1996. Other Asian countries accounted for over 65 percent of Japan's factory export business. Business in countries damaged by the currency crisis fell sharply. Indonesian business fell from $1.6 billion in 1996 to $430 million in 1997. Thailand business dropped from $2.3 billion to $460 million. Forecasts for 1998 anticipated the plant-export business would fall below $10 billion. To counter the decline, the Japanese government established a $46 billion loan program for the Export-Import Bank of Japan and lowered the interest rate on official development assistance to 0.75 percent with a forty year repayment period. Japan's plant-engineering industry employed 140,000 engineering workers. Japan's top eighteen trading houses, which accounted for 30.3 percent of total plant exports, say plant exports fell by 39.6 percent in 1997. The global market for plant engineering was $126.8 billion in 1996 [Fukuda 1998].

2.3.2. Japanese Imports

The leading categories of imports for 1997 were machinery and equipment (11.5 trillion Yen), fossil fuels (7.5 trillion Yen), foods (5.6 trillion Yen), and raw materials (3.6 trillion Yen). The import of aircraft caused a strong 9.5 percent growth in the import of machinery and equipment. In contrast, machinery and transportation, which accounted for 8.6 percent of imports, were down by 17.5 percent in 1997. The Middle East, with U.A.E. and Saudi Arabia leading, was the primary source of Japan's crude oil.

Table 2-6 Top Ten Commodity of Imports (1992-1996)
(hundred million Yen)

Rank	1992	1993	1994	1995	1996
1st	Crude Oil 38,121	Crude Oil 31,392	Crude Oil 28,253	Crude Oil 28,201	Crude Oil 36,442
2nd	Textile Articles 19,391	Textile Articles 18,331	Textile Articles 20,794	Textile Articles 23,128	Textile Articles 27,885
3rd	Fish and Shellfish 15,923	Fish and Shellfish 15,513	Fish and Shellfish 16,231	Fish and Shellfish 16,313	Business Machines 20,527
4th	Timber 9,679	Timber 11,308	Timber 9,985	Business Machines 14,814	Fish and Shellfish 18,134
5th	LNG 9,245	LNG 7,981	Business Machines 9,248	Semi-conductors and Other Electronic Components 11,509	Semi-conductors and Other Electronic Components 14,406
6th	Non-ferrous Metals 8,913	Business Machines 7,606	Nonferrous Metals 8,083	Nonferrous Metals 10,169	Automobile 11,521
7th	Meat 8,188	Meat 7,576	Meat 7,938	Automobile 9,585	Timber 10,402
8th	Petroleum Products 8,010	Noferrous Metals 7.485	Semi-conductors and Other Electronic Components 7,474	Timber 9,437	Meat 10,191
9th	Coal 7,702	Coal 6,606	Motor Vehicles 7,378	Meat 9,054	Nonferrous Metals 10,090
10th	Business Machines 7,604	Semi-conductors and Other Electronic Components 5,983	LNG 7,090	LNG 7,204	LNG 9,402

Source: *Kanzei Nempo* (Annual Report of Customs and Tariff), Customs and Tariff
Bureau, Ministry of Finance, June 30, 1997.

Japan is the world's largest net importer of agricultural products and the largest market for U.S. farm products. As shown in Table 2-6, crude oil, fish, timber, and meat were among Japan's top ten imports. Japan's Food Agency controls all facets of production, marketing, and trade of rice, wheat, and barley. Rice, wheat, and barley are subject to stringent import control by the agency.

Table 2-7 Japan's Electronics Imports by Category (1988-1996) (million Yen)

Year	Consumer electronic equipment	Industrial electronic equipment	Electronic components and devices	Total
1988	97,773	435,328	700,199	1,233,300
1989	145,399	586,169	959,969	1,691,536
1990	113,090	692,421	1,195,075	2,000,585
1991	135,680	687,422	1,286,074	2,109,176
1992	156,041	633,332	1,256,228	2,045,601
1993	172,457	697,674	1,307,877	2,178,008
1994	238,499	880,887	1,589,417	2,708,803
1995	333,273	1,433,037	2,198,527	3,964,837
1996	410,437	1,999,329	2,879,063	5,288,829

Source: Ministry of Finance, Japan

Food products accounted for 12 percent of imports in 1997. Raw materials accounted for 13 percent of imports, with lumber accounting for over 1 trillion of the 3.6 trillion Yen of imports.

Japanese 1996 imports rose 20.4 percent in Yen value over 1995. This growth was not due to increased volume, but to price increases caused by the devaluation of the Yen (unit volume rose only 2.3 percent compared to a 17.7 percent increase in import value). Prior to the Yen's devaluation, the unit value of import in Yen fell through 1995 as lower quality and less-added-value imports rose. As shown in Table 2-7, imports of consumer electronics rose 23.2 percent, while those of industrial electronic equipment grew 39.5 percent, and those of electronic components and devices advanced 31.0 percent. Japanese firms have increasingly imported ICs and microassemblies from subsidiaries across Asia. The top ten electrical products are shown in Table 2-8.

2.4. Public Spending and Restructuring

After the 1989 collapse, the 1990s have seen continued review of Japanese industrial policy. Slow economic growth has continued since the beginning of the decade, with little evidence of recovery. The government has initiated annual stimulus packages since 1992 in an effort to resuscitate the business climate and help stimulate the international economy in Asia. Stimulus packages include:

- August 1992:   10.7 trillion Yen
- April 1993:   13.2 trillion Yen
- September 1993:   6 trillion Yen
- February 1994:   15.25 trillion Yen
- April 1995:   7 trillion Yen
- September 1995:   14.22 trillion Yen
- April 1998:   16 trillion Yen

For 1999, public works programs were increased another 7 percent over 1998. According to Clyde Prestowitz, former U.S. trade negotiator, "The only way for any of the stimulus packages to become effective is to be coupled with far-reaching deregulation and antitrust efforts that would remove the administrative bottlenecks and anti-competitive business practices that make everything from houses, to airline tickets, to taxi rides, to photo film two or three times as expensive in Japan as in other industrialised countries."

2.4.1. Impetus for Restructuring

Three major problems were affecting Japan's industrial competitiveness: (1) as shown in Table 2-9, Japan had the highest industry cost structure, due to strict regulations covering the financial, public, energy and distribution sectors of the economy, (2) As shown in Table 2-10, Japan's taxation system was not competitive with other countries in attracting foreign companies, and (3) Japan

Table 2-8 Growth in Import of Electrical Equipment as Capital Goods
(Top Ten Products)

	1990 (% Share)	1996 (% Share)	Growth %	Contribution %
1) ICs and microassemblies	387.7 (20.5)	1,280.2 (33.9)	238.1	48.9
2) Electrical apparatus for line telephony	82.0 (4.4)	286.2 (7.6)	248.9	11.1
3) Television receivers	24.0 (1.3)	161.5 (4.3)	574.3	7.5
4) Parts of radio equipment	60.2 (3.3)	166.1 (4.4)	176.0	5.7
5) Transmission apparatus for radio telephony	31.2 (1.7)	137.1 (3.6)	339.3	5.7
6) Rectifiers and transformers	82.4 (4.5)	184.2 (4.9)	123.6	5.5
7) Insulated conductors and cables	51.7 (2.8)	140.4 (3.7)	171.6	4.8
8) Other electrical equipment	111.9 (6.1)	198.1 (5.2)	77.0	4.7
9) Motors and generators	58.8 (3.2)	116.9 (3.1)	98.8	3.1
10) Semiconductors & photosensitive devices.	59.0 (3.2)	111.5 (3.0)	89.0	2.8
Subtotal of top 10	939.9 (50.9)	2,782.2 (73.6)	87.2	31.9
Total of electrical equipment	8,097.4 (100.0)	3,779.2 (100.0)	17.9	17.9

Note: Growth and contribution are comparisons of two time periods, 1990 and 1996.

Source: JETRO White Paper on International Trade 1997: Global trade in the era of information communications. Japan External Trade Organization, 1997, p. 49.

Table 2-9 Japan's High-Cost Business Structure (Japan=100 in table)

		US	Germany	South Korea	China	Singapore
Energy costs	Petroleum products;	67	117	152	54	53
	Industrial power rates;	77	81	44	71	38
	Industrial water rates	--	--	191	39	374
Transportation/ telecommunication costs	Railway cargo	61	67	24	10	--
	Coastal shipping	131	--	40	22	--
	Port transport costs	90	--	47	--	53
	Int'l. airfreight rates	55	73	98	36	80
	Custom clearance costs	--	--	24	--	20
	Local telephone charges	97	155	52	14	29
	Long distance tel. charges	48	65	23	5	--
	Exclusive int'l. line	88	101	95	--	85
	Exclusive local line	32	185	123	87	--
Fund-raising costs	Bond issuing cost	86	--	--	--	--
Land develop costs	Plant site	71	62	54	--	--
	Commercialized	11	24	28	9	38
	Office rental	55	52	56	135	70
Personnel expenses	Personnel expenses	73	135	31	--	31
Tax	Corporate income tax	82	100	65	66	54

Source: Japanese government publicity "Six reforms" (June 1997)

Table 2-10 Japan's High Corporate Tax Rate

As of March 1998 (% of revenues)	Japan	Japan after 1998	US	UK	Germany	France
National tax	33.48	31.08	31.75	31.00	34.18	33.33
Local tax	16.50	15.28	9.30	0.00	15.61	0.00
Effective tax	49.98	46.36	41.05	31.00	49.79	33.33

Source: *Action Plan for Economic Structure Reform*, Industrial Policy Bureau, Ministry of International Trade and Industry, March 1998, p. 12.

lacked advanced technology in aerospace, information, and biotechnology that was needed for future new product creation and low-cost production.

The government's 1998 restructuring package included proposed tax cuts and tax credits for housing and equipment investments, welfare payments, public works projects, improved telecommunications infrastructure, help for small- and medium-sized businesses, and assistance for Asian countries. It is intended that tax relief will stimulate domestic demand and that assistance to Asian countries will help stabilize their financial systems and expand their economies and related demand for Japanese exports. Japan's economic planning agency estimated that the GNP would still grow by only 1.6 percent in 1998 as unemployment reached a record 4.2 percent [Takahashi 1998].

Governmental deregulation is intended to address the high cost structure of Japan in six areas: energy, transportation and telecommunications, fund raising, land development, personnel expenses, and taxes. A government analysis compared Japan's costs to those of other countries in 1997. The government is committed to creating an attractive business environment comparable to those foreign countries through deregulation and tax reform. Unfortunately, actions to implement changes prior to 2001 have been limited. Deregulation of the Antimonopoly Act to allow regulated companies to enter new business areas is scheduled for 1999. In the area of taxation, measures are planned to reduce the effective rates of national and local taxes imposed on enterprises, which now are higher than in other countries.

Japan has also identified fifteen new industrial sectors in which jobs and growth are expected. Targeted industries include medical care and welfare, quality of life and culture, information and telecommunications, new manufacturing technologies, distribution and logistics, environment, business support services, oceanographic technology, biotechnology, improvement in urban surroundings, aviation and space, new energy and energy conservation, human resources, economic globalization, and housing.

2.5. Program to Stimulate Foreign Imports into Japan

The Japanese government has initiated reforms in its economic structure intended to foster new industries, create a business environment attractive by international standards, harmonize taxes and social security costs, and reduce the public's economic burden. Specific policies include the promotion of imports by shortening and simplifying import procedures, coordinating international standards for goods, and providing information concerning imports. The goal of Japan's import expansion program is to maintain and encourage economic and trading partnerships with other countries, and to expand the volume and variety of imports entering Japan.

2.5.1. Expanding Japan's Imports

The Japan External Trade Organization (JETRO) (1) dispatches specialists overseas to search for products, including products of developing countries, that will appeal to Japanese consumers, and (2) provides strategic support to foreign companies already engaged in exporting to Japan and brings foreign sellers and Japanese buyers together through trade fairs and exhibitions. JETRO's Business Support Centers in Tokyo and other Japanese cities assist foreign businesspeople entering the Japanese market.

2.5.2. Fostering Industrial Cooperation

JETRO fosters technology transfers and manufacturing investment to and from Japan. JETRO has sixteen Centers for Industrial and Technological Cooperation that promote cooperation between industrial organizations in Japan and North America, Europe, and Oceania. JETRO provides information on the investment climates in countries worldwide and offers a variety of services tailored to companies considering direct investment in Japan. Its Business Global Partnership program encourages cooperative partnerships between Japanese industries and their counterparts overseas.

2.6. Openness for Trade and Investment

JETRO promotes small business, foreign currency-generating industries, and transfer of energy-related technologies to Japan. It provides market information on Japan providing specialized know-how for improving product quality, and developing human resources of trade promotion organizations in developing countries. JETRO also supports technological cooperation in energy and environmental protection by dispatching specialists to consult with industry leaders in developing nations.

2.6.1. International Communication

JETRO works to ease international friction and create mutually rewarding economic and trade relations by participating in projects to help leaders broaden their understanding of Japan. It publishes information about Japan's economic trends, trade, industry, technology, market development, and business practices for dissemination to public and private entities.

2.6.2 Inward Direct Investment

Table 2-11 shows that the number of foreign capital companies investing in Japan had been increasing at about 10 percent annually. With the devaluation of the Yen and the deregulation of markets, many foreign companies took the occasion to increase the amount of their investments and take a stronger position in the Japanese market. North American firms continued to lead the way in 1996 with 431 cases, amounting to 244.5 billion Yen. European firms were second in number, with 353 cases, amounting to 220.2 billion Yen. Asian-based firms were third, with 297 cases, totalling 137.2 billion Yen in 1996. Firms from Latin America had even made a move to increase their investment in 1996, with 51 cases valued at 65.6 million Yen.

2.6.3. Foreign Access Zones

The government of Japan, operating under a special law enacted in 1992, has established a network of Foreign Access Zones (FAZ) around the country to attract increased imports and foreign investment. A total of twenty-two prefectures had established FAZs by March 1997. Each FAZ had easy access to a port or airport with cargo-handling facilities for rapid distribution and logistic support. Each zone required a designated land area, with approval of

Table 2-11 Inward Direct Investments in Japan (hundred million Yen)

Fiscal Year	1993		1994		1995		1996	
	Number of Cases	*Amt	Number of Cases	*Amt	Number of Cases	*Amt	Number of Cases	*Amt
Total	1,072	3,586	1,135	4,327	1,272	3,697	1,304	7,707
Japan: Foreign capital	161	393	163	320	162	233	151	1,000
North America	332	1,262	362	1,981	475	1,786	431	2,445
Europe	294	1,199	313	1,586	330	1,274	353	2,202
Asia	236	539	233	271	242	247	297	1,372
Middle East	3	3	6	0	4	0	7	20
Latin America	30	173	32	136	27	141	51	656
Africa	3	5	-	-	7	10	-	-
Oceania	13	12	26	33	25	5	14	14

Note: Inward direct investments in Japan are investments by foreign capital companies.
Source: *Conditions of Inward and Abroad Direct Investment*, International Finance Bureau, Ministry of Finance, June 1997. (*Amt=Amount)

plans by the ministry in charge. The Minister of International Trade and Industry, the Minister of Transport, the Minister of Agriculture, Forestry and Fisheries, and the Minister of Home Affairs provided guidelines for the act. Corporations organized to conduct promotional activities and approach companies to encourage them to operate in the zones.

2.7. Outward Expansion of Offshore Facilities

To stay competitive in world markets, Japanese electronics manufacturers are cutting back on their labor forces, downsizing plants and facilities, reducing product lines, and reducing the frequency of product model changes. In addition, manufacturers continued to move their production facilities overseas. Of four hundred overseas production bases owned by Japanese electronics makers, 60 percent were for consumer electronics plants and 40 percent for industrial products. The major shift in facilities began in 1985, with a move into Southeast Asia.

The largest number of Japan's Asian plants was located in the ASEAN (Association of Southeastern Asian Nations) countries of Malaysia, Singapore, Thailand and, most recently, Indonesia. The move to ASEAN countries was in response to Japan's need to restructure global operations in light of declining production advantages in Japan, the Yen's appreciation, trade friction, and the unification of regional markets. These plants provided low and medium-priced products and components for the electronics industry. By 1991, of the 136 billion Yen in imported consumer electronics goods in Japan, 86 percent came from Asia, with only 8 percent from the United States, 5 percent from Europe, and 1 percent from elsewhere.

A significant number of Japanese companies continued to expand plants in Southeast Asia and China during 1993. However, investment was shifting to China. After having invested nearly $30 billion in Malaysia, for example, Japanese investment there fell to only $1.5 billion in 1992. The number of Japanese investment applications fell from forty-six to thirty-five for the periods from January to April of 1992 and 1993, respectively. Singapore maintained its high level of investment from Japan with nearly $3.5 billion in investments for 1993 (*Newsweek* 1993). In an annual survey of the Electronics Industry Association of Japan (EIAJ) membership in June 1997, 180 of the 422 responding companies reported 1,221 production facilities outside of Japan, an 8.1 percent increase over 1996 (Table 2-12). The People's Republic of China is the most popular country for direct investment, with 234 Japanese electronics factories due to its robust economy and abundant labor resources. North America had 210 production facilities. Another 147 were in Europe and twenty-five in Latin America.

Table 2-12 Offshore Facilities of EIAJ Members (as of June 30, 1997)

	Production	R&D	Finance	Regional head offices
Asia	829	34	16	53
Europe	147	26	27	42
North America	210	67	33	56
Latin America	25	0	0	2
Africa	4	0	0	0
Oceania	6	5	3	0
Total	1,221	132	79	153

Source: "A Year of Brisk Activity in Direct Overseas Investment," *EIAJ Review*, March 1998, p. 4.

Forty-seven companies had 132 R&D facilities abroad, of which sixty-seven were in North America, thirty-four in Asia, twenty-six in Europe, and five in Oceania. Twenty members also claimed seventy-nine financial facilities. North America had thirty-three facilities, Europe twenty-seven, Asia sixteen, and Oceania three. Fifty-nine firms report 124 regional head offices as of June 1997–fifty-six in North America, fifty-three in Asia, forty-two in Europe and two in Latin America. Japanese firms were moving headquarters closer to their target markets to improve efficiency and tailor production to the needs of local markets. As a result, the number of regional headquarters increased by twenty-nine over the prior year.

As shown in Table 2-13, Japanese firms continued to invest abroad, though the number of cases fell in 1996 to 2501, with a value of 540.9 billion Yen. North America, Asia, and Europe were the three most active

Table 2-13 Direct Investments Abroad by Region (1993-1996)
(hundred million Yen)

Fiscal Year	1993		1994		1995		1996	
	Number of Cases	*Amt.	Number of Cases	*Amt.	Number of Cases	*Amt.	Number of Cases	*Amt.
Total	3,488	41,514	2,478	42,808	2,863	49,568	2,501	54,094
North America	953	17,591	534	18,525	551	22,394	638	25,933
Latin America	327	3,889	303	5,499	300	3,741	257	5,008
Asia	1,478	7,672	1,305	10,084	1,629	11,921	1,233	13,083
Middle East	16	251	12	303	3	148	10	268
Europe	494	9,204	221	6,525	260	8,281	241	8,305
Africa	52	630	35	366	37	367	41	485
Oceania	68	2,275	68	1,507	83	2,716	81	1,011

Source: *Conditions of Inward and Abroad Direct Investment*, International Finance Bureau, Ministry of Finance, June 1997. (*Amt. =Amount)

Table 2-14 M&A Data by Market after 1991

Year	IN-IN	IN-OUT	OUT-IN	OUT-OUT	Total
1991	310	292	18	18	638
1992	254	179	29	21	483
1993	236	108	24	29	397
1994	250	187	33	35	505
1995	255	208	33	35	531
1996	321	226	31	43	621
1997	455	216	51	33	755

Source: *1997 M&A (Out-In) a Record High with a Backdrop of the 'Big Bang', Investment News,*
No. 13, 1998, p. 3.

investment locations for Japanese firms. North America accounted for nearly half of Japan's foreign investment. Investments amounted to nearly 2.6 trillion Yen in 638 cases, suggesting that capital intensive investments averaged over $300 million each. Asian investments were down in 1996 to 1233 cases with a recorded value of 1.3 trillion Yen, averaging $8 million per investment. European direct investments amounted to 830 billion Yen in 241 cases, or an average of about $25 million per investment. Latin America and Oceania followed with Japanese investments of 500 billion Yen and 101 billion Yen, respectively, in 257 and 81 cases, averaging about $16 million per investment for Latin America and $10 million each for Oceania.

2.7.1. Mergers and Acquisitions

Table 2-14 shows that the number of mergers and acquisitions (m&a) in 1997 reached 755, exceeding the previous high of 754 in 1990. The first m&a boom in Japan occurred in the period from 1988 to 1991. The second boom started in 1994 and had yet to reach its peak in 1998. The various forms of m&a can be seen in the following exhibit. There were 455 mergers and acquisitions between Japanese-capital companies within Japan (IN-IN). There were 216 m&a transactions by Japanese companies purchasing foreign-capital companies overseas (IN-OUT), and 33 transactions between Japanese capital companies outside of Japan (OUT-OUT).

2.8. WTO Requirements to Deregulate the Financial Services Industry

In spite of the increased openness of Japanese trade and investment, the World Trade Organization's requirements for open market access in financial areas had yet to be achieved. Japan's timetable had been established in 1997 as follows:

1997:   • Securities houses can handle consumer payments for their clients.

1998:	• Companies and individuals can handle foreign-exchange transactions without government authorization.
	• Banks can sell their own investment trusts over the counter.
	• Ban on financial holding companies lifted.
	• Firms can become securities brokerages without government licenses.
1999:	• Securities houses can expand asset management services. Securities houses will set commissions on securities trading of any size.
	• Market-value method will be used for marketable securities.
	• There will be no barriers for banks, trust banks, and securities houses to enter each other's markets.
2001:	• Banks will issue straight bonds.
	• Banks and securities houses can enter insurance sector.
Future:	• A new financial services law would govern banking, securities, and insurance sectors.

2.8.1. Foreign Exchange Law

As of April 1, 1998, many of the legal barriers that shielded and shut off Japan's financial markets for decades were lifted. The foreign-exchange statute, renamed the Foreign Exchange and Foreign Trade Law, opens foreign-exchange transactions. Non-financial companies, including convenience stores and ticket outlets, can participate. The new law allows:

- any individual or enterprise to be in the foreign-exchange business;
- residents to have accounts in any currency at foreign banks without prior authorization from the finance minister;
- residents and non-residents to make transactions in foreign currencies;
- domestic investors to make securities transactions with foreign securities companies without prior notification or authorization.

The new law allows businesses and individuals to manage their own money, which means that there will be more competition between financial institutions. Services are becoming better and cheaper. Japanese firms engaged in global operations will be able to rationalize their foreign currency transactions and net transactions without having to pay the cost of currency transactions.

2.9. U.S.-Japan Relations

2.9.1. Japan's Trade Balance with the USA

Trade relations between the U.S. and Japan have always been strained by a continuing imbalance, as shown in Table 2-15. Japan has been accused of being restrictive in its trade practices and continues to be pressured to deregulate its markets. Japan's high cost of doing business makes it difficult to sustain profitable operations in Japan. Japan's primary exports to the U.S. in 1996 included capital equipment (62.1 percent of exports), consumer products (24.1 percent), and industrial supplies (10.6 percent). In contrast, U.S. exports to Japan in 1996 were composed of foodstuffs and raw materials (22.5 percent), manufactured products (47.5 percent), and other electronics and equipment (25 percent).

The top ten U.S. manufactured exports to Japan in 1996 included:
1. $5.9 billion in computers (31.5 percent of imports),
2. $6.1 billion in integrated circuits (51.6 percent of imports),
3. $2.1 billion in aircraft (81.3 percent of imports),
4. $2.4 billion worth of engines, turbines, pumps, and boilers (59.8 percent of imports),
5. $1.9 billion in medical instruments (57.1 percent of imports)
6. $2.6 billion in measuring and testing equipment (57.1 percent of imports),
7. $3.8 billion worth of microcomputers and processors (43.7 percent),
8. $1.8 billion in telecommunications equipment (42.4 percent of imports),
9. $654 million in radioactive chemicals (61.6 percent of imports),
10. $324 million in blood-related agents (65.5 percent of imports).

The U.S. was the world leader in most of these product areas and held a major share of imported products, as shown above. Japan was very interested in keeping abreast of the most advanced technologies available from the United States. In fact, the U.S. accounted for about 70 percent of the value of Japan's technology imports in 1995 [*Handy Facts on U.S.-Japan Economic Relations* 1998].

Table 2-15 Japan's Trade Balance

Year	Japan Imports (total)	Japan Exports (total)	World Trade Balance	US Export to Japan	US Imports from Japan	US Trade Balance
1992	233	340	107	52	96	44
1993	241	361	120	55	105	50
1994	275	396	121	63	118	55
1995	336	443	107	75	121	46
1996	351	412	62	80	112	32

Source: *Handy Facts on U.S.-Japan Economic Relations: 1998.* Japan External Trade Organization, p. 16.

Chapter 3

History of the Japanese Electronics Industry

3.1. Basis for Growth after World War II

Japan experienced social upheaval almost overnight twice in her modern history — in 1868, when she relinquished her three hundred year-old isolation from the outside world, and in 1945, when the nation surrendered to the allied forces. The event that culminated in the restoration of the emperor to the formal pinnacle of the nation in 1868 (Meiji restoration) was a process that spanned several years, but the sudden influx of western technology shocked the populace, which had had little access to information from abroad for many generations. The national experience that Japan underwent with the Meiji restoration is unique in world history in the suddenness with which it brought about a reversal in social and cultural values, from medieval conformity to a scramble for things from the West. It brought both blessings and traumas to the nation that have lasted to this day.

One of the most visible blessings was a nationwide aspiration to learning. Graduation from schools was conductive to a higher living standard. School enrollment quickly rose; by 1900 the rate of enrollment in primary school reached more than 90 percent of the entry-age population. In particular, education in western science, technology, and culture at the imperial universities, with the University of Tokyo at their pinnacle, meant a promising career and high personal prestige. This rush to learning cultivated a corps of factory workers and engineers who were extremely competent in adopting Western technology and quick to accommodate themselves to emerging new technologies.

The trauma was a conviction that anything new should come from the West, not from within. Western knowledge became a trade commodity, even at the individual level. Being knowledgeable about Western technology was a fast way to win social respect. Intense competition for access to Western knowledge frequently produced embarrassing situations. Once imported by someone, any new idea fast diffused in the society, and the identity of inventors was lost. If the idea was for a small, easy-to-manufacture commodity, an industry producing similar commodities sprang up almost overnight.

A value system inclined toward the West accompanied a nationalist backlash, and the odd combination of extreme nationalism and modern military technology was fermented by the political instability and famine of the 1930s. Aroused by Japan's outward expansion and her military build-up, the Western powers tried to curb Japan's naval capability, in particular, by allocating Japan a smaller quota for naval ship tonnage than that of other nations. The imperial navy dedicated its resources to offsetting the tonnage handicap by improving the quality of its strike power. It had close collaboration from the engineering faculty of the University of Tokyo and other imperial universities, and developed the Japanese shipbuilding industry to world class. While the navy built advanced battleships, it also turned to the then-novel concept of an air strike force. Some critics later speculated that the design of zero fighter planes came from literature published in the West in earlier years. Whatever the truth, the realization of ideas required a certain ingenuity on the part of its implementers. This argument, pitting the generators of the idea against the toil of the implementers, remains a recurring issue today.

By the time World War II was over, the industrial infrastructure of Japan was severely damaged. However, the nation was left with a work force and engineers who were well acquainted with modern technology. The wartime mobilization effort involved even those of middle-high school age; almost everyone who was in the mid- and low-teens around 1945 had factory experience. Some worked on the absurd project of producing bamboo spears, but many worked in assembly plants for airplanes, communications equipment, and weaponry. Although not all of these drafted youngsters went into post-war factories, their experience allowed many of them to jump into assembly plants when the time came. Also, many were motivated to enter science and engineering schools by their experience in the wartime factories.

The end of World War II produced another awakening in the Japanese to the power of Western technology. In the Meiji era the flow of technology was mostly from Britain, Germany, and France, but this time America was where industry turned. America defeated Japan with her technology. The American occupation forces shocked the populace with their sturdy, fast-moving Jeeps, gigantic B-29s and, more than anything else, their material affluence. The scramble to acquire American technology began and an America-first value system developed. For instance, study in America on the Fulbright Fellowship Program became an avenue for ambitious young people to pursue fast-track careers. Indeed, the Fulbright Program helped produce a corps of competent post-war leaders in the government, industry, business, and cultural sectors.

If the war in the Pacific theater was the result of Japan being overtaken by surreal nationalism, what occurred after the war also showed a Japan gripped by another abstract concept — namely, pacifism. Although Douglas MacArthur, Supreme Commander of the occupation forces, imposed the ninth clause in the Japanese post-war constitution, which prohibited the possession of armed forces as a means to settle international conflicts, pacifism has been a widespread popular sentiment. Military technology became taboo, in particular, in academia. Pacifism allowed Japan to pour all of her resources into the development of civilian technology. It was accompanied, however, by a social system and psyche developed before the war. The government retained its authority and its power in directing business and industrial practices. This meant national regimentation under powerful bureaucrats. The society in general also retained its respect for vertical social order. The wartime respect for generals and admirals was replaced by respect for corporate presidents. Just as advancement in military ranks was the way to gain respect in militaristic society, now moving up corporate ladders occupied many. In place of the young and ambitious army officers who drove Japan toward a precipitous course, Japanese companies now had drivers in the middle management ranks who often steered the companies on aggressive business courses.

The combination of human resources, government power, freedom from military development, mutual competition in a regimented society, and yearning for new technology made post-war Japan an extremely efficient machine for commercial implementation of technological ideas. Moreover, the highly charged energy at the grass roots level provided fertile ground for rapid development of Japanese industry after the war. For example, Takahashi (1993) provided an account that showed the interplay between this grass-roots enthusiasm, incentive given by the American occupation forces, the Japanese government, and the established Japanese companies in the reconstruction of the radio industry after the war. Radio sets were among the first commodities that were in urgent demand right after the war. The occupation forces needed them to broadcast public education programs that denounced wartime propaganda by the Japanese military. The populace needed them for access to information and entertainment, which has been scarce during the war years. The radios owned by the people were in such poor shape that the pronouncement of surrender by the emperor on August 15, 1945, was masked by crackling noise and barely audible. As early as the end of 1945, the occupation forces issued orders for four million radio sets. This spurred manufacturing by established firms as well as small entrepreneurs. The manufacture of components such as resistors and capacitors was left mostly to small entrepreneurial firms that hired housewives and dependent youngsters. Components were purchased by wholesale merchants who served as middlemen between the component manufacturers and the set assemblers. A significant fact was that radio sets were assembled not only by established firms but by a number of amateur assemblers. The first author recalls that even older primary school children who aspired to be scientists joined the excitement, using a motley assortment of hand-made components and items released from former Japanese military depots.

Takahashi (1993) recounted that already in 1948, the production volume of radios, (about 800,000) and the number of radio-equipped households exceeded the highest prewar levels. During 1949 – 1950 a severe economic recession was precipitated by the tough monetary policy needed to curb inflation, conceived by an American banker and consultant to the occupation policy board, Joseph Dodge. Although many of the assembling firms were wiped out, the radio component industry survived this recession and even thrived, thanks to the unabated demand from amateur radio assemblers and the efficient distribution net established by the wholesale merchants.

The Korean War of 1950 – 1953 uplifted the economy, then another recession set in. During this period the landscape of the radio industry underwent considerable change. In 1952, the NHK's monopoly of broadcasting was broken when the Ministry of Communications allocated frequency bands to private broadcasters. In 1953 the first television broadcast was aired from NHK. What might be called grass-roots democracy in the radio industry eroded with the growth of the industry. The export of portable radio sets had achieved a considerable volume already by 1955. The export of television components, such as capacitors, resistors, and speakers, also rose to a visible level in that year. Both the growth of the home market and access to foreign markets required high-quality components, compelling component suppliers to contract manufacturing licenses from abroad. By 1957 more than fifty-seven Japanese companies had been licensed to manufacture television components by RCA, Western Electric, EMI, and Philips. Assembling advanced radio and television sets fell into the hands of technologically competent large firms, such as Hitachi and Toshiba. Only those component manufacturers who lined up with large assembling houses to receive technical guidance and steady purchase orders survived. The government provided quality control guidance by issuing codes and standards, and guidance came trickling down from large firms to component suppliers.

By the early 1960s the hierarchical structure of the industry and the machinery for exporting high-quality products were established. Other capital-intensive industrial sectors, such as electric-power generation, chemicals, and materials processing, achieved substantial recovery from the devastation of World War II. Japan, Inc., was ready to jump into the semiconductor age. It was, however, not the government-led machinery that first set eyes on one of the important inventions of the twentieth century.

3.2. Entrepreneurial Thrusts in Consumer Electronics

Post-war Japanese industrial development was due to two mechanisms; government-led machinery, and the grass-roots scramble for new technologies. The former system has been magnified in the eyes of Western observers; however, the latter has played equal or more significant roles, particularly in the development of consumer electronics. These two mechanisms often proved incompatible; the demand for resources from small start-up businesses irritated the government, which wanted to preserve precious resources for big companies. In what follows we attempt to illustrate grass-roots energy by providing example episodes.

3.2.1. SONY and Transistor Radios

Tokyo Telecommunications Research Laboratories was founded on May 7, 1946, with twenty strong employees. It was this small company in ruined Tokyo that paid serious attention to the transistor invented in 1947 at Bell Laboratories of AT&T. Akio Morita's account [Morita, 1986] tells how a small company could get on a rapid growth path to becoming an industrial leader, SONY, using the transistor radio as its first booster. It also details the archetypal penchant for new ideas and new things common to Japanese people.

Close to the end of World War II, Morita, a physics graduate of Osaka University and a naval engineering officer, worked on an imperial navy project on heat-seeking devices. The project team had a civilian participant, Masaru Ibuka, the owner of a small company that had developed a submarine-detection device for the navy. Shortly after the war, Morita helped Ibuka to gear up his company in a corner of the devastated downtown of Tokyo. Their expertise led them to choose short-wave radio adapters as the first product line. Radios were urgently needed commodities, but Ibuka chose to specialize in the short-wave adapter to carve out a market niche ignored by large companies. Their next line of products was motors and pickups for phonographs, second only to radio sets in satisfying peoples' thirst for entertainment.

The company could begin turning profits on the manufacture of adapters and phonograph components, but Ibuka and Morita did not want to stop there. Ibuka envisioned new consumer products, and set his eyes on recording equipment. While they explored the feasibility of wire recorders, thanks to some luck and personal connections, the company received an order for a large broadcast mixing unit from NHK. Broadcasting technology was at that time supervised by the American forces, whose technical staff introduced advanced communications technology to NHK. The trips to NHK gave Ibuka a chance to spot an American-made tape recorder in the NHK laboratory. He and his company started attempts to manufacture tapes in place of an earlier recording medium made of metal wires. Plastic tapes were not yet available, and they had no expertise in depositing magnetic powders on the tape. They worked with cellophane tapes and paper tapes, resorting finally to manually coating magnetic material on these test tapes. The results were a shambles. However, at the same time they developed other hardware components, which were ready just as true plastic tapes became available to them.

After an initial period of blindly searching for sales, Morita found markets for tape recorders first in the Japan Supreme Court, where stenographers were in short supply, then in schools, where teaching English became mandatory. The growth of their business gave the company the capacity to look further for novel products. In 1952 Ibuka made a trip to the U.S. to view state of the art of tape recorder usage there, but found that tape recorders were already more abundant in Japanese schools than in American schools. Like many other Japanese entrepreneurs, Ibuka and Morita were keen to learn of any new technological developments in the U.S., and they had already read the news about the invention of the transistor at Bell Labs. While Ibuka was disappointed by the lack of interest in tape recorders in the U.S., he learned more about the transistor — in particular, the forthcoming availability of licenses to transistor manufacturers.

A year later, in 1953, Morita went to the U.S. to sign a licensing agreement with Western Electric, a manufacturing arm of AT&T. He received advice from Western Electric that, if he wanted to use the transistor in consumer electronics, the hearing aid was its only possible application — its operative frequency range fit only the range of audible sound. Aiming at the bigger market of radio sets, Ibuka, Morita, and their staff attempted to make the transistor workable in a radio frequency range. The key to success seemed to be in the choice of proper dopants. At least one of the research team members was aware of an earlier attempt at phosphorus doping at Bell Labs, terminated when it did not produce promising results.

Phosphorus doping was doggedly pursued by the Ibuka-Morita research team. They eventually achieved dramatically improved product yield. During their trial-and-error research, one sample showed a peculiar behavior that prompted researcher Leo Esaki to develop the tunnel diode theory. Esaki won the Nobel Prize in 1973 for this feat. Some more time was needed for the company to refine the phosphorus-doped transistor, package it in a compact radio set, and work out a mass-production scheme. The company produced its first transistor radio in 1955, and two years later, in 1957, its first 'pocketable' radio. About this time the company adopted a new name, SONY, which helped identify the company with an image of miniaturized radios.

The story of Morita included an episode that indicates the government's initial apathy toward the arrival of the semiconductor age. Ibuka and Morita went to the Ministry of International Trade and Industry (MITI), applying for a permit to pay the licensing fee to Western Electric. The dollar was scarce at that time, and overseas payments were under the strict control of the government. They met enormous resistance from MITI officials concerned about the well-being of big established companies. A tiny unknown item like the transistor did not deserve precious dollars in the eyes of MITI officials. Ibuka's eloquent persuasion finally got him a permit after six months of effort.

Indeed, even in the late 1960s the white papers on science and technology issued annually by the Agency of Science and Technology mentioned semiconductor technology, but apparently not with much enthusiasm. The white paper of 1969 included nuclear energy, space technology, and ocean technology in the primary list of new technological frontiers. It added information processing, bioscience, and materials research as areas of future interest.

3.2.2. The Calculator War

What finally ushered Japan into the age of integrated circuits was not the powerful MITI, but the life-and-death competition among calculator manufacturers, known as the calculator war, which raged in the early 1970s [Aida, 1995-1996]. In the peak year of 1971 there were about fifty competitors, large and small. Out of that competition Sharp and Casio emerged as survivors. What happened during this period illustrates several prominent characteristics of Japanese consumer electronics development: interactions with American forerunners, sweatshop toil to improve product quality, cut-throat internal competition, cooperation between small and big companies, and the government as a provider of business infrastructure.

Modernization of Japanese business management was one of MITI's primary concerns in the post-war era. In 1952, Law for Promotion of Streamlining Business Operations (*Kigyo Gohrika Sokushin Hou*) was promulgated. The next year, Committee for Streamlining Business Operations (*Sangyo Gohrika Shingikai*), an advisory arm for MITI, proposed the promotion of computer (or calculator) usage in business management. Responding to this call a number of small and large manufacturing companies stepped up the effort to develop calculators. Most of their products were inexpensive calculators based on manual relay mechanisms. Engineer-entrepreneurs in small companies were eager to apply new driving mechanisms to calculators. By the late 1960s, desk-top calculators driven by electric motors and mechanical relays became popular in business offices. Those calculators, however, were noisy, heavy, and slow. It was obvious that quieter and faster calculators would capture a big market if they were priced correctly.

In 1957, Kashio Manufacturing Company (later Casio) developed a calculator based on electro-mechanical relays. The product was a commercial success, but due to its relatively large physical size, customers were limited to large business firms. In 1961, Nihon Calculator (later Busicom), imported a calculator from the U.K. This calculator, "Anita Mark 8", developed by Samlock Comptmeter, used vacuum tubes, but its size was suitable for the office desk. Big and small companies set their sights on this product, and soon worked hard to replace vacuum tubes with transistors and incorporate other new features. Sharp and Canon were first in "transistorizating" the calculator.

The year 1964 was called "the beginning of *Dentaku* (electronic calculator) era." At the Tokyo Business Show that year, Sharp, Canon, Ohi-Denki, and SONY displayed desk-top calculators that drew enormous attention. SONY's calculator was the most advanced in size and design, and used an integrated circuit (IC). However, actual production didn't begin until three years later.

Development of integrated circuits in the U.S. was closely watched by the Japanese calculator manufacturers. In 1971, Yoshio Kojima of Busicom shocked his competitors by releasing an IC-based pocket calculator. To develop this model he contacted Intel and MOS Tech — both then spin-out companies, the former from Fairchild and the latter from Texas Instruments — and asked them to utilize Busicom's circuits on several silicon chips. Even better, MOS Tech put the circuits on one chip, enabling the calculator to be shrunk to pocket size. Its high price, a little less than twice the starting salary of a college graduate, left this first one-chip calculator out of reach of general consumers; nonetheless, Busicom's calculator triggered the race toward IC-based calculators.

A far more decisive jolt came from Texas Instruments Japan at almost the same time as Busicom's move. TI Japan put calculator software on a chip developed by Gary Boon and Michael Cochran of TI. The chip, the first of the TMS-1000 series, was released in 1971. This chip reduced manufacturing calculators to a simple assembling job, eliminating the need for professional wiring and expensive capital investment. Small companies sprang up, some operating in family houses. According to an estimate [Aida, 1996], the number of assemblers exceeded fifty at the peak of the calculator war. Most small companies were original equipment manufacturer (OEM) assemblers for bigger companies. In this repetition of the radio boom of the late 1940s, the scale of competition was far greater, due to increased market capacity in Japan's high-growth period. The sales volume of TMS-1000 chips grew rapidly, from fifty thousand to three hundred thousand per month by the time TI Japan built a mass production facility in Saitama Prefecture in 1973.

The performance of calculators, particularly those in the lowest price range, was no longer a primary competitive issue; new display features became the crucial factor. The key price issues were automated assembly, chip price, and sales network. The Busicom calculator of 1971 cost 89,800 Yen; in 1972, Casio released the Casio-Mini for 12,800 Yens, reflecting the tough price competition that arose in a short period. The companies at the extreme ends of the spectrum began dropping out — family-size companies were wiped out as OEM sponsors withdrew their orders to cut costs; big companies like Hitachi, Toshiba, Mitsubishi Electric, Fujitsu, and NEC withdrew because their businesses in other sectors gave them less incentive to stay in a tough race. Instead, they became chip vendors for calculator manufacturers. This move was very similar to that of TI in the U.S., which first developed an IC-based calculator but chose to remain in the role of chip vendor. The chip for the Casio-Mini was developed jointly by Casio and Hitachi, which already had the capability of designing and manufacturing integrated circuits by the time Casio brought its six-digit calculator to Hitachi. NEC sold MOS chips to Sharp. The calculator war closed when a final drop-out in 1978 left Casio and Sharp as winners. Casio survived due to the huge success of the Casio-Mini and the sales net established among stationary retailers.   Sharp focused on automated manufacturing and relocated assembly lines to lower labor-cost regions.

Many episodes show MITI's ambiguous roles in the development of the calculator industry [Aida, 1995-1996]. MITI advised Sharp to focus on consumer electronics when large-scale computers became the focus of research and development in large electrical companies. As described above, Busicom served as a technology pioneer several times in the history of Japanese calculator market. In one of its moves, Busicom developed an innovative calculator that incorporated ferromagnetic core memory elements, announcing it in 1966. A patent of the memory device was bought from an Italian inventor. The Business Machines Association, an arm of MITI, and MITI, itself tried to discourage Busicom from announcing the product, saying that the calculator market was already covered by Canon and Sharp

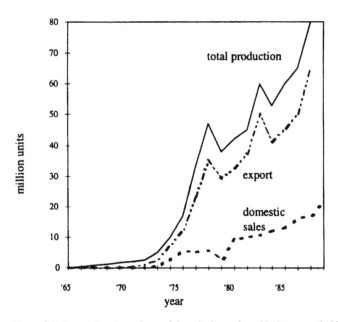

Figure 3.1 Growth in sales volume of the calculators from 1965 to around 1985
[Tanimoto, 1980]

and a new product would disrupt the market balance. Busicom did not take the advice. Neither did the product topple the market. Rather than encouraging or discouraging agents in specific technological developments, MITI, the Ministry of Finance, and other government ministries played important roles through financing and technology standardization in establishing an industrial and business infrastructure that allowed the calculator industry to grow to become export-intensive. Figure 3.1 [Tanimoto, 1980] shows the growth in calculator sales volume from 1965 to around 1985. The export share was dominant after 1975.

The role of the calculator war in the development of Japan's IC capability is illustrated by the diagrams in Figure 3.2. From 1971 to 1973, the demand from the calculator sector topped all other demands. Subsequently, the demand for mainframe and medium-sized computers dominated, but in 1977, calculators recovered the top share. Calculators helped not only to develop the Japanese IC industry, but also acted as a catalysis for microprocessor development at Intel.

Again Busicom served as a door opener. In 1968 Busicom's Kojima was visited by Robert Noyce, who then made a tour in Japan selling Intel's new invention, DRAM. Impressed by Noyce, Kojima chose Intel as Busicom's partner to develop a new integrated circuit for calculators. The next year, Busicom sent three engineers to Intel. At about the same time, North American Rockwell was making substantial profits by selling QT-8D chips to Sharp. Intel initially assumed that their work would reproduce the Rockwell-Sharp venture. However, what one of Busicom's engineers, Masatoshi Shima, disclosed to Intel was a far more ambitious idea: using ten Rockwell-Sharp venture. However, what one of Busicom's engineers, Masatoshi Shima, disclosed to Intel was a far more ambitious idea: using ten

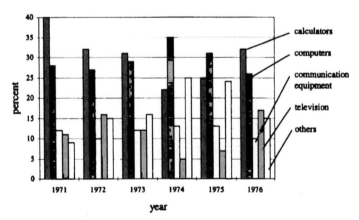

Figure 3.2 Role of the calculator war in the development of Japan's IC capability

chips. Ted Hoff of Intel jumped on this proposal from Busicom to work on his favorite idea, developing general-purpose calculator chips. Hoff conceived an architecture that reduced the number of chips to four. Moreover, the central processing unit chip was unprecedented, in that the chip's function could be defined by programming, a feature that differentiated microprocessor chips from custom-made ICs. Intel completed its first microprocessor 4004 in 1971. The rest of the story of Intel's emergence as a leading microprocessor vendor is well known. Later, in the heat of the calculator war, Busicom went bankrupt, but Shima went on as a logic designer at Intel, then at Zilog.

In the early phase of new technology introduction, some aberrations from the supposed Japanese norm occurred. The expectation of lifetime employment was often disregarded. Key technical players moved from company to company, from the government laboratories to private companies. Companies plucked engineers from their competitors. The import of American semiconductors grew to 36% of market share by 1974, as Figure 3.3 shows [Aida, 1995-1996]. In particular, MOS LSI chips, for which Japanese chip vendors lagged well behind that of American vendors, were aggressively purchased by Japanese calculator manufacturers over protests from their home-grown vendors.

3.2.3. Craftsmen in Town Shops

For the electronics industry to develop into a full-fledged national industrial sector, it needs an infrastructure that extends over a diverse spectrum of endeavors. Today, more than five hundred industries support the operation of a semiconductor plant, starting with silicon crystal growth, through intermediate processing involving (among other steps), mask production, photolithography, diffusion, chemical and physical etching, pure water generator, and dice cutting, to testing and packaging. The path to such

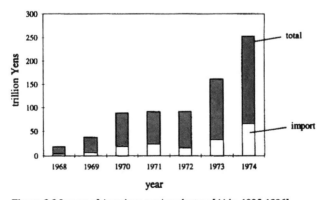

Figure 3.3 Import of American semiconductors [Aida, 1995-1996]

infrastructure development was bumpier than what is implied by today's image of Japan, Inc.   In the early 1960s Japan's government-industry machinery was still concerned with so-called heavy-large industries, such as steel making, ship building, and nuclear power development. Any attempt to work on microscopic devices such as transistors and IC chips were met with disdain from mainstream bureaucrats and engineers, who maintained a macho image about anything called "industry."   Those pioneers who participated in the initial phase of electronics industry development had to seek cooperation among craftsmen in the small machine shops in towns. Some of these shops later grew into companies that specialize in certain manufacturing technologies and capture large shares in world market niches [Aida, 1995-1996].

One of these, Disco, specializes in dice cutting and has captured 75% of the world market in dicing machines. Disco began as a manufacturer of grinder tools in Kure, Hiroshima, where the imperial navy had its largest shipbuilding capability in pre-war years. After the war, its founder, Mitsuo Sekiya, came to Tokyo and purchased a small company that had a technology for sintering grinder particles with clay. He replaced clay with phenol resin, and succeeded in manufacturing 1mm-thick grinder discs. Sekiya's grinder discs were met with great demand from the utility companies, which needed them to manufacture watt-meters for homes. A home-building boom in post-war Japan was the first boost for Disco. The name of the company became well known to those people who needed precision cutting. The next boost came from the manufacturers of fountain pens, who needed 0.14mm-thick grinder discs to cut out pen tips. Disco's expertise in reducing the thickness of the grinder disc was greatly advanced by the time the boost came from transistor manufacturers. The required thickness of the grinder disc, which was reduced to less than 0.1mm, seemed too demanding, but Disco eventually came up with a 70 μm thick disc. As in many other early-phase endeavors, there were personal interactions between the user and the grinder disc developer, through several cycles of trial manufacturing and evaluation. The first big user of Disco's grinder was Hitachi, who applied it to cut silicon wafers.

In 1968 Disco began marketing 70 micrometer-thick grinder discs. Marketing to Japanese companies, and later to American firms, was met with skepticism at first. Even where grinder discs were purchased, the blame for any trouble in final products was often placed on the disc. To be free of such concerns, Disco embarked on the development of its own dicing machine. After a series of debacles, Disco produced a dicing machine and took it to Semicon West 1975. The machine used a micro-controller and, of course, the Disco disc. The machine's capability in precision cutting without replacing the grinder disc was a sensation. TI Japan first purchased Disco's dicers, and its success triggered worldwide orders from TI.   Other semiconductor manufacturers followed.

A similar ascent to world-class status was achieved by Mitsui Hitec, which now has a 30% share in the world-wide supply of leadframes. Mitsui Hitec, unrelated to the wealthy clan Mitsui, started as a three-man company specializing in the manufacture of punching dies. The initial boost to the company came when founder, Takaaki Mitsui, succeeded in adapting tungsten carbide as a die material around 1950. The tungsten carbide die did not wear quickly, and proved cost effective for manufacturers of electric motors. The company grew as it received orders not only from Japanese electric companies but also in growing numbers from American companies. In 1968 Takaaki Mitsui traveled to the U.S., and was first acquainted with photo-etched leadframes for ICs. Mitsui was convinced that a leadframe of equal quality and geometric precision could be formed by punching. The key was the design of a punching tool. Mitsui constructed it by stacking a number of thin tungsten carbide blades, a method that proved effective in drastically reducing the production cost to about one-tenth of the cost of photo-etching. His aggressive sales in the U.S. finally caught the attention of TI in 1969, which placed an initial order amounting to 500 million Yen. In the 1970s Motorola, Fairchild, National Semiconductor, and others followed suit. The company's sales grew to 30 billion Yen by 1990.

Shinkawa Ltd. is one of the leading manufacturers of wirebonding machines, with a 40% worldwide share as of 1992. Founded in 1959 with ten employees, Shinkawa's initial venture was to serve as an intermediate inspecting agent between transistor manufacturers and soldering vendors (the latter performed electrolysis plating of solder on three pins of the transistor). By around 1964 Shinkawa began manufacturing bonding machines upon request from their transistor manufacturing patrons. This first-generation machine was operated manually. Transistor manufacturers hired a corps of young female workers who patiently performed bonding using the manual operating machines.

With the advent of the IC, the number of I/O pins increased dramatically, and so did the demand for factory workers. The industry experienced a shortage of workers in around 1970. The situation was similar in the U.S. American manufacturers chose to shift their assembly lines to Southeast Asia, where low-wage factory hands were available. Low-cost ICs began arriving in Japan. In particular, the calculator assemblers chose imported ICs over homemade ones. Japanese IC manufacturers tried to cope with this challenge by automating wirebonding. A clue to automation came from an American invention, ball-bonding, the development of which was claimed by both Kuric & Sofa and Fairchild. Japanese transistor manufacturers embarked on the development of automated bonding machines around this invention. Shinkawa, as a specialized bonding machine manufacturer, was naturally in this race.

Shinkawa's engineer, Mikiya Yamazaki, set his eyes on Intel's microprocessor 4004, after reading that it was a novel device in one of the

1971 issues of *Nikkei Electronics*. Yamazaki eventually developed an automated bonding machine using Intel 8008. The first microprocessor-based bonding machine was released from Shinkawa in 1973. This machine succeeded in improving the efficiency of assembly lines almost four-fold. Shinkawa has continued to relentlessly pursue automation, incorporating a number of features that have reduced the need for human intervention in the assembly line. K&S followed suit developing microprocessor-based bonding machines. Today, the market is shared primarily by Shinkawa and K&S.

The list of craftsmen-entrepreneurs does not end here; but extends over the whole industrial pyramid which supports the production of electronic products. Although there are certain variations in the foundation and growth-path among different companies, including those which are now well-known, such as Kyocera, the key to their growth was the incessant improvement of manufacturing skills.

What enabled fledgling companies to embark on their growth paths was the patron-apprentice relationship with engineers of large established companies. The patron often made mind-boggling demands on the apprentice in cost-cutting, shorter delivery cycle, and escalating product quality. The patron evaluates the product of the apprentice, and throws harsh criticism on the apprentice. The apprentice endured as in the feudal era; in exchange for the hardship, he learned cutting-edge technical information from the patron who had greater access to advanced information. For the apprentice, a good relationship with the patron means a secure flow of orders and a reputation with a well-known large company. During his work in Hitachi, the first author often heard from component vendors such phrases as, "We owe Hitachi-san for our ascendancy to our present-day status," and "We were trained really well by Hitachi-san."

The patron-apprentice relationship is also maintained within the company group, which has a pyramid structure with a final product assembler at the top. From mid-level companies with such hierarchies there emerged worldwide suppliers of microelectronic materials, components, fabrication equipment, testing equipment, and others.

Craftsmen labored not only in small town shops, but also in the factories and laboratories of large companies. They worked together with pioneering engineers who held higher college degrees than they did. These teams of engineers and craftsmen were initially outside the mainstream, regarded as mavericks who risked their careers in uncertain products like transistors. Japanese manufacturing companies had been enjoying a supply of high school graduates from the agricultural areas; for such employees, a job at a well-known large company had won their respect within their close-knit communities. Large companies recruit top-class graduates, and give them the training and opportunities to step up the corporate ladder, although handicapped with respect to their peers with college degrees. In the established sectors, such as heavy electrical machinery, it has been extremely difficult for the high school graduate to go beyond a certain point in the managerial pyramid — generally, the level of the shop floor technical manager. In newer sectors, such as microelectronics, the barrier was relatively low, opening up a wide avenue for aspiring high school graduates. The rapid expansion of the microelectronic business also helped to provide opportunities for high school grads. These were hardworking people, and stayed with the same factory while their college-educated colleagues moved to other posts with the expansion of business. Some of them grasped shop floor know-how better than anyone else. In each factory there was at least one such tech master who grew up with the development of the microelectronics technology, virtually ran the factory, and exuded authority sometimes beyond the factory manager. Some advanced beyond the rank of division manager, and enjoyed an international reputation for their technical expertise.

3.3. The Government and Large Companies

A patron-apprentice relationship has long been forged on the national scale between the government and the leading large companies. Although in the initial phase of its development the transistor evaded the serious attention of government bureaucrats and top corporate management, developments in the U.S. woke the sleeping giant. The news which drew attention from Japan, Inc. originated with the big American companies, AT&T and IBM. The companies' size and fame won the attention of the centralized national machinery. By this time, there was already a corps of people in the industry and in government laboratories who had advanced technical expertise in the production and applications of transistors without being in the limelight in their home institutions. The chief accomplishment of those early pioneers of the Japanese electronics industry was the improvement of product quality and yield, from germanium transistors to silicon mesa transistors to MOS transistors. Governmental sponsorship helped to bring attention to those pioneers and encouraged the research and development effort to meet stricter quality control targets.

3.3.1. NTT and the DEX Projects

The invention of the transistor at Bell Laboratories in 1947 was motivated by the desire to produce reliable high-speed switches for telephone exchanges. However, it did not lead to immediate abandonment of crossbar electromechanical switches, even at AT&T. Rather, few engineers embraced the idea of electronic switching because of the then advanced technological state of crossbar switches [Fransman, 1995]. More than ten years later, in 1960, Bell Labs built the first electronic switching system based on the stored program concept (SPC). In 1965, the first commercial electronic switch based on SPC, No.1 ESS, was produced by Western Electric and AT&T.

Nippon Telephone and Telegraph (NTT) had been the sole domestic telecommunications carrier in Japan until the mid 1980s. It was a public company under the jurisdiction of the Ministry of Communications until 1985, then was partially privatized. NTT is still the largest company in the world in terms of stock value. In 1964, the year Bell Labs announced AT&T's forthcoming No.1 ESS, NTT launched an electronic switching system project, dubbed the DEX-1 project. Following common practice for national projects, NTT called in the companies under its long-time patronage: NEC, Fujitsu, Hitachi, and Oki. The technical hub of the project was NTT's Electrical Communications Laboratories (ECL). The DEX-1 project was completed in 1966 and followed by the DEX-2 (1966 - 69) and DEX-21 (1969 - 71). Finally in 1972, the first commercial digital switch, DC10, was built by NEC. These projects involved a scheme called 'space-division', indicating fixed-line connection during the time of usage, which was soon replaced by the time-division scheme based on the concept of time-sharing.

The commercial benefits of the DEX projects, therefore, were marginal for participating companies, as far as the telephone switches were concerned. However, the real impact followed semiconductor manufacturing. NTT requested the participating companies to subject all integrated circuits to be used in telephone switches to far sterner acceleration tests than those adopted by NASA. The reliability of made-in-Japan IC chips was raised to world-class level by leverages from NTT, the big patron [Aida, 1995-1996].

The process of transition to the time-division scheme was a typical example illustrating the interplay between the government and industry. Like its counterparts in the U.S. and European countries, NTT initially held a conservative attitude toward the time-division scheme, although it conducted an exploratory project in the mid-1960s. The companies under NTT's wing naturally went along with NTT's view.

Meanwhile in the world market, Northern Telecom spearheaded the move to time-division by selling its DMS10 to an independent carrier in Florida in 1977. By that time, companies were notified of heightened U.S. interest in time-division by their business outposts. While casting a hesitant look at NTT, NEC initiated development on time division; so did others. NTT engineers were eventually alerted to the arrival of the new technology at the International Switching Symposium in Kyoto in December 1976. In 1977 NTT organized a project for the time division, again calling in NEC, Fujitsu, Hitachi, and Oki. These companies had already spent their own R&D money and developed some technical expertise in this technology. The project produced the D70 local switch and the D60 toll switch, which went into operation in 1982 and 1983, respectively. Already prior to the completion of the project, NEC, Fujitsu, and Hitachi had sold their versions of the time-division switch overseas.

Switches by Japanese manufacturers, however, have hardly become a source of trade friction in the international market, unlike automobiles, semiconductors, and consumer electronics. A major reason is the policy of the patron, whose primary concern has been updating the technology in Japan. As telephone exchange technology was merging with computer technology, NTT needed to keep its hand in the development of operating systems software to ensure the compatibility of equipment procured from different suppliers. The operating system is naturally oriented toward the Japanese telecommunication system, which has different historical and cultural backgrounds from those in other countries. The cost of re-tailoring the operating system has prevented companies from upscaling the overseas marketing.

As described so far the government's role in technology development may seem relatively passive. In reality, however, the government plays dual roles; it sometimes acts as a spearhead for a new direction. In government laboratories, in particular, attention was directed to research projects considered novel and exotic by researchers elsewhere. Transistors were studied at MITI's Electrical Research Laboratory right after the news from Bell Labs. NTT's ECL has also served as probe for new technologies. Another motivation for government to look in new directions came from the advisory panels composed mainly of academics from top-notch universities, who were inclined toward novel ideas. The decision whether to pursue new technologies or behave conservatively depends on the available resources. As the nation grew rich, the pendulum tended to swing to technologically aggressive paths. Since 1991, NTT has been leading the R&D for asynchronous transfer mode (ATM) switching, which is critical to the era of multimedia communications. The project involves not only Japanese companies such as NEC, Fujitsu, Hitachi, Toshiba, Oki, and Mitsubishi Electric, but also Northern Telecom and AT&T.

3.3.2. MITI and the VLSI Project

Interest in large-scale computing in Japan was largely confined to the scientific community until the late 1950s. Because of several developments in Japan and abroad that occurred in the next few years, MITI and the leading companies came to realize the strategic importance of computers. One such development was the research interest in digital telephone exchange switches. Another was the development of a ticketing and reservation system for Shinkansen bullet trains, which began operation at the time of the Tokyo Olympic games in 1964. IBM's System 360, introduced in 1964, created a sensation that reached beyond the U.S. The 360 had several then novel features: the use of hybrid integrated circuits, software compatibility with existing computers, and time-sharing (in the later versions). It solidified the position of IBM as the technological leader among computer manufacturers.

When Japan awakened to the dawn of the computer age, the industry realized a wide technological gap yawned between them and the U.S. manufacturers. In the early 1960s the leading Japanese companies, one after another, struck technological agreements with American companies in order to learn the technology: Hitachi with RCA, Mitsubishi Electric with TRW, NEC with Honeywell, Oki with Sperry Rand, and Toshiba with General Electric. The exception was Fujitsu, which started its own development effort earlier in the 1950s, getting some funding from MITI, part of MITI's budget for exploratory R&D. Fujitsu's strategy in embarking on computer manufacturing earlier was to impress NTT with its technological prowess in computers, then pass NEC in the ranking of NTT suppliers [Fransman, 1995]. Indeed, many academics and graduate students in Japanese universities in the mid 1960s had their first experience in computing with Fujitusu Machines, known as FACOM. Fujitsu's early entry into computer manufacturing helped push the company to the top rank in the Japanese computer market, where it has stayed to this day.

In response to the IBM System 360, MITI organized a "very high speed computer system project," which continued from 1966 to 1972. MITI's Electrical Research Laboratory served as technological coordinator; Hitachi, Fujitsu, and NEC developed the hardware; and Toshiba, Mitsubishi Electric, and Oki developed peripheral equipment. Although the companies eventually developed commercial computers based on their experience with this project, skepticism pervaded the engineering staffs from competing companies. It often happened in national projects like this that MITI's endorsement was used to persuade corporate management to support in-house R&D, but technological exchanges among companies were minimal. The fund from MITI was just a token of its endorsement compared with the total budget borne by the companies.

In 1970 IBM introduced the System 370, which shook off some competitors in the computer market. Between 1970 and 1972 GE, RCA, and TRW decided to exit. In the face of overwhelming dominance by IBM, MITI acted to realign the Japanese companies. The disappearance of American partners from the computer market forced Japanese companies to choose one of the three paths — independent development, seeking a new partner, or exiting outright. Fujitsu had Amdahl, an IBM-compatible maker, as partner. Hitachi decided to develop computers on its own. With MITI's coordination, Fujitsu and Hitachi were paired to develop IBM-compatible computers known as the M-series. NEC chose to pursue a non-IBM-compatible line. Toshiba, Mitsubishi Electric, and Oki eventually quit R&D on mainframes.

It is difficult to estimate the impact of MITI's funding of two projects, the Mainframe Computer Project (1972 - 1976) and the Pattern Information Processing System Project (1971 - 1980), on the subsequent emergence of three Japanese companies as serious competitors to IBM. Besides the DEX projects, NTT launched a project (called DIPS project) which focused on development of IBM-compatible computers for telephone exchange systems. Fransman [1995] shed light on the role of the DIPS project, but again, its real impact is ambiguous. The government's contribution is difficult to assess in monetary terms because of vague calculations of R&D expenditures on the part of the private sector. The situation remains the same today. Often, several projects run in parallel in the company; while they overlap in technical content, some are related to the national project, but the majority are aimed at the company's own commercial products. What was obvious, however was the effectiveness of MITI's role in buying time for the domestic computer industry to develop. Purchasers of large-scale computers were national organizations such as universities, government laboratories, the national railway, and corporate giants such as steel manufacturers and banks. Those purchasers were most susceptible to pressure from MITI's buy-Japan initiatives.

Other factors also worked to reduce the potential of foreign competitors, in particular IBM, in the Japanese computer market. Large users in the private sector and domestic suppliers were linked through the banks. Those links, dating back to pre-war years, are known as *Keiretsu*. The institutional force worked to purchase capital equipment, such as mainframes, from a supplier in the same *Keiretsu*. Psychological kinship between personnel in the user and supplier companies in the same *Keiretsu* group facilitated close aftercare and continuous updating of operating systems in response to emerging user needs.

These factors worked to strengthen the competitiveness of Japanese manufacturers. The share of IBM in the Japanese market was as high as 80% in the 1950s, but shrank to less than 30% by the late 1970s. From 1976 to 1979, Hitachi increased computer sales by 50%, NEC almost 75%, and by 1979, Fujitsu took over Japanese sales leadership from IBM.

MITI's concern with the domestic market and the institutional ties between users and suppliers, shaped Japanese competitiveness in the computer market. Japanese mainframes became competitive in the domestic market, but the growth of exports was not rapid enough to cause concern in the American and European markets. A far greater impact on the world market was produced by the Japanese development of memory chips. For this, MITI's Very-Large Scale Integration (VLSI) Project (1976 - 1980) became well known, and was given credit later by many foreign market and technology analysts.

The VLSI Project was motivated by the news of IBM's Future System document, which forecast the arrival of one-megabit memory chips by the early 1980s. Concerned about its widening distance from the cutting edge of technology, NTT's ECL drew up a plan, "Approach to VLSI," in 1974. MITI independently launched the VLSI Project in 1976. It was agreed between NTT and MITI that the former would focus on communications applications and the latter on VLSI for general-purpose computers. The VLSI project was run by the VLSI Cooperative Research Association, which included Fujitsu, NEC, Hitachi, Toshiba, and Mitsubishi Electric, and under its wing there were three laboratories: Computer Laboratory, NEC-Toshiba Information Systems Laboratory, and VLSI Joint Laboratory. Engineers from the participating companies socialized and mutually probed their competitors' thinking. The alumni of MITI's VLSI Project now recount that they never discussed actual manufacturing; instead, their mutual interest was in learning the fundamentals of VLSI technology [Aida, 1995-1996].

The primary concern of MITI's VLSI project was the development of electron-beam equipment. This was a compromise based on the interests of the participating companies and the loftiness required for the national project. Photolithography was, and still is, a major method for fabricating integrated circuits. The companies had already committed their own resources to developing photolithographic technology, so they had no incentive to disclose their expertise to rivals. Electron-beam lithography was regarded as next-generation technology, which would be used in manufacturing lines in some unknown future. Taking up a then futuristic topic such as the electron-beam technology on top of the agenda justified the expenditure on the part of MITI.

The biggest impact of the VLSI Project was made on the companies who were not directly involved in the project but received orders for equipment from the VLSI Cooperative Research Association — Canon and Nikon. Canon was an early starter in supplying photolithography equipment to chip makers. Nikon was invited to join the equipment market through the VLSI Project, focusing on the step-and-repeat aligner (stepper), and eventually emerged as the world's leading supplier of aligners by the mid-1980s. The VLSI Project set targets of lithographic spatial resolution and alignment accuracy, with eyes well on the future. The approach involved close collaboration between the chip makers and the equipment suppliers. The patron-apprentice relationship worked effectively to remove an American supplier, GCA, which had enjoyed a high share of the stepper market before the emergence of Japanese competitors.

Like the VLSI Project, MITI's subsequent projects — the Supercomputer Project (1981 - 1989) and the Fifth Generation Project (1982 - 1991) — were aimed at technologies beyond those already commercialized elsewhere. The Supercomputer Project included R&D on superconductors and molecular beam epitaxy (MBE). As the infrastructure for Japanese chip making and computer manufacturing grew to maturity, large fall out from the national project became less likely. Meanwhile, the dominance of made-in-Japan memory chips in the 1980s overlapped with the VLSI project in the eyes of foreign, particularly American, observers. Many of them feared that ensuing projects might irreparably damage their country's competitiveness. For instance, the Fifth Generation Project raised emotional concern about the national safety of the U.S. [Feigenbaum and McCorduck, 1983], although the argument seemed partly designed to draw U.S. government funds to research on artificial intelligence.

Government funding of R&D in the 1980s served two functions for private companies. One was to support small research groups whose assignment was to assess the feasibility of novel ideas that drew the attention of the worldwide scientific community. Another was to manufacture and deliver advanced equipment to government laboratories. The situation has not changed much in the 1990s. However, difficulty arose for MITI with the diversification of electronics technology, the maturation of the Japanese industry, and the uncertainty of emerging technologies. It becomes increasingly hard to find glamorous projects that would justify expenditure and wholehearted commitment from corporate giants.

3.4. Trade Friction

It has often been pointed out that R&D concentration on DRAM chips was a wise strategy adopted by Japanese manufacturers in view of the more rapid growth of commercial demand for DRAM than for logic chips. But pursuing DRAM manufacturing was more a natural consequence for the Japanese semiconductor industry than a deliberately calculated intention. Designing DRAM chips involves less software development, an area in which Japan lagged well behind that of the U.S. Besides, Japanese software development was focused on the domestic market. On the other hand, DRAM chips have the highest transistor density among different chip types. The growth of transistor density has faithfully followed Moore's law, which predicts a doubling of the number of transistors on the chip every eighteen months. For an industry with an efficient hierarchical machinery best suited for manufacturing volume items, DRAM was a natural fit, and Moore's law served as a guide for further technology development rather than mere record of past achievement [Schaller, 1997].

The simplicity of DRAM design allowed new entrants in the market other than the established electrical companies. Due to the huge capital investment required for start-up companies, the entrants were mostly subsidiaries of big companies in steel making, ship building, and other traditional manufacturing sectors. The production of DRAM chips soon surpassed domestic demand, and the export drive began.

Silicon cycles have been repeated every few years since the appearance of DRAM chips in 1970. The share of made-in-Japan chips in the world DRAM market was nil in 1970, then grew to 80 percent in 1988. The turning point for U.S. suppliers came around 1983, when they delayed preparing for the transition from 16K-bit to 64K-bit DRAM in the middle of a bottoming silicon cycle. The transition from 64K-bit to 256K-bit started in 1985, but by then the Japanese suppliers had established an overwhelming lead in the DRAM market. American suppliers such as Motorola, Intel, and Advanced Microsystems exited the market one after another. By the mid-to-late 1980s only Texas Instruments and Micron Technologies remained in the market.

The flood of made-in-Japan DRAM and other memory chips into the U.S. market brought political repercussions. Dumping charges were brought before the U.S. Department of Commerce and the International Trade Committee. The U.S. Semiconductor Industry Association urged the U.S. Trade Representative to investigate unfair trade practices in the Japanese market. The Japanese government sought to work out a political solution. In July 1986, an agreement was reached by the U.S. and Japanese governments on three points: first, the share of imports in the Japanese market would be raised from the then current 10 percent to 20 percent within five years; second, Japan would establish an organization to promote imports; and third, Japanese products would be subjected to a cost surveillance system. Japanese chip suppliers were obliged to report the manufacturing cost of DRAM and EPROM chips to the U.S. Department of Commerce in every fiscal quarter. A dumping incident also occurred in the European market; the EC responded, introducing the minimum cost line in 1990.

Through the late 1970s and 1980s the leading Japanese chip suppliers began manufacturing operations in the U.S. NEC Electronics USA acquired a plant in Silicon Valley in 1978, then built a VLSI plant in Roseville, California, in 1984. Hitachi Semiconductor USA built a plant in Dallas, Texas, in 1978, Fujitsu Microelectronics in California (1979) and Oregon (1988), Toshiba Semiconductor USA in California (1980), and Mitsubishi Semiconductor USA in North Carolina (1983). In Europe, NEC Ireland built a plant in 1974, NEC Semiconductor UK in 1975, Hitachi Semiconductor Europe in Germany in 1975, Fujitsu Microelectronics Ireland in 1975, Toshiba Semiconductor GmbH in Germany in 1983. Those overseas plants were initially designed to perform post processing and packaging, but intensifying trade friction and the rising Yen forced expansion of the operation to cover all fabrication steps. By the late 1980s almost all of them became full-fledged manufacturing plants.

Overseas production, directives from MITI to buy made-in-America chips, and internal pressure by corporate management on procurement departments tipped the trade balance the other way only a little. Tensions have lingered, and the issue of trade imbalance remains current. The reasons for the ineffectiveness of these measures lies in the nature of the DRAM market. First, DRAM chips on the circuit board outnumber microprocessor and other logic processing chips by a factor of as many as fifty, so that the game cannot be won unless foreign suppliers upscale DRAM production to match the number gap. To date, only Korean companies have decided to enter the DRAM market.

Second, the organization and culture of the industry make a difference. Competitiveness within the DRAM market is rooted solely in the timing of capital investment to update and expand fabrication facilities. If the timing of production expansion coincides with the bottom of a silicon cycle, the company suffers a huge loss. Any huge loss, however, can be absorbed by Japanese DRAM suppliers, most of whom are manufacturers of a broad range of industrial products. Covering the loss with the income from other sectors, they can wait until the next upturn of the market. In the Japanese corporate atmosphere, expansionists are likely to win the ear of top management, and would not be blamed if the investment proved unprofitable one or the other in the short term. In any event, the value of this investment has been proven in the long run as the DRAM market expanded.

Third, the production of a new generation of DRAM chips involves cooperation among a wide spectrum of materials and equipment suppliers. Easy communications fostered by long-term relationships between the chip maker and the support companies within the Japanese corporate culture are essential to quick-start production of a new generation of chips. Hence, chip makers manufacture the latest generation of chips at domestic plants, and shift the production of older chips to overseas plants. Soon after their introduction, new generation chips outnumber older ones, contributing to unabated flows of DRAM chips from Japan.

The Japanese dominance of the DRAM market stirred concern among U.S. microprocessor manufacturers that microprocessors would be the next target of Japan, Inc. Indeed, rhetoric about the threat from Japan was used by Andy Grove in his pep talk to the employees of Intel in 1984 [Malone, 1995]. Intel's effort to improve product yield by stricter quality control might have checked the Japanese venture into the microprocessor market, but only marginally. The microprocessor is a product whose value and performance depend heavily on the software. Moreover, the development of operating systems for microprocessors is like exploring the unknown using a vision encompassing a wide spectrum of applications.

The Japanese were good at the development of computer programs for well-defined applications, as illustrated by the many programs they developed for mainframes operating in finance, business, administration, science, and engineering. By the same token, the Japanese were good at the design of micro-controllers that fit specific applications with much less circuit complexity than microprocessors. Indeed, the production of micro-controller chips soared in the 1980s, as they were used in diverse consumer products such as rice cookers, microwave ovens, baths, washing machines, refrigerators, air-conditioners, toilets, cameras, and videos often with the commercial phrases "fuzzy-logic controlled" or "neuro-fuzzy controlled." As with software for mainframes, software for consumer products did not establish a market of their own beyond the domestic realm. Rather, a majority of those micro-controller chips went abroad, hidden in various consumer products.

Also working against a large-scale Japanese foray into the microprocessor market was the copyright attached to the microcode, the central software engine for the microprocessor. The 1980s saw a series of lawsuits brought by American manufacturers, notably Intel and Motorola against Japanese companies. The suits made news, but some time later, faded out as settlements were worked out in ways that discouraged the Japanese companies from attempting a large presence in the microprocessor market.

3.5. Synopsis of History to the End of the 1980s

Two distinct driving forces have shaped the Japanese electronics industry: the popular force and the institutional force. The popular force is an amalgam of craftsmen-entrepreneurs who are ubiquitous at all levels of Japanese industry. They are keen to learn new scientific discoveries and quick to identify their commercial feasibility, and they serve as virtual drivers of technological development, particularly in the initial phase of development. Some researchers performed their own experiments, trying to confirm what they learned from the news about the invention of the transistor at Bell Laboratories. Reports of the invention of integrated circuits at Texas Instruments and Fairchild Semiconductor also spurted a flurry of in-house efforts to reproduce the invention, as did news about the advent of the microprocessor. Most of those hastily arranged early efforts were primitive, reflecting ignorance of the key factors involved in the original invention, such as materials selection and the control of materials purity. Those efforts, however, served as an introductory self-learning process, and smoothed the way for the introduction of original technologies through licensing. Once the details of an original invention were brought to the licensee, they were rapidly absorbed; often, to the chagrin of the original inventor, improvements were quick to follow. Such improvements were largely made on manufacturing techniques, and their key aspects were difficult to document, because they were achieved mostly through trial and error. The Japanese are not used to documenting the details of what they do; this applies, in particular, to those craftsmen in small town shops who formed the backbone of the manufacturing system. Besides, when things worked out successfully, theoretical interpretations of the success were not worth attention in the competitive atmosphere. Eventually, once knowledge was exported to Japan, only commercial implementations, often accompanied by superb improvements, came from Japan. The trade imbalance in knowledge (a deficit for Japan) and goods (an excess for Japan) has not been corrected to this day.

Institutional inertia made the government and the top management of large companies initially cool to the arrival of new technologies that had not been proven elsewhere. They were eventually awakened by developments in the key technological sectors in the U.S. — namely, telecommunications and computers — and by the surge in exports of electronic components. By that time, the technological infrastructure in the private sector was already developed to some extent. The roles played by the big government projects, such as DEX and VLSI, were two-fold. First, the government agencies imposed high reliability standards for electronic components, like a master goading the private company disciples to come up with ever-better results. Government-sponsored projects also promoted standardization of commercial components. Second, the government decided who participated in which aspects of the projects, assigning companies to specific technological sectors for which they became chiefly responsible. These arrangements were designed to avoid overheated competition in the domestic market, as well as to avoid leaving any essential technological areas unattended. The result was an infrastructure for manufacturing microelectronic chips that became self-sufficient in the domestic market, covering every phase from the production of silicon ingots to packaging to inspection. With such government encouragement, the Japanese electronics industry grew into an efficient export machine. This competence in export, however, was largely limited to volume-intensive components like DRAM chips. Japanese consumer electronics products, which have traditionally been strong in the world market, also became increasingly competitive due to the infrastructure build-up aided, in part, by government projects.

These effects of government projects were, however, by-products in terms of the primary objectives defined on paper. The government projects aimed at the development of large systems for domestic usage, such as central telephone exchange systems and mainframe computers. Export of those large systems grew somewhat, but never became a serious threat to competitors in other countries. The technology development for those systems, however, received undue attention from abroad. In particular, supercomputers won high visibility in trade negotiations between the U.S. and Japan. Also, financing the projects with public funds led the Japanese to assume that government agencies and national universities should purchase the products of these national projects. Such buy-Japan practices received heated criticism from the U.S. government and competitors.

The effectiveness of government projects has lessened as the industrial infrastructure has matured. Through institutional inertia, MITI and other government agencies have continued to launch national projects. To persuade the Finance Ministry to support a project, the project is often focused on technological fronts regarded as cutting-edge by academia, but lacking any immediate commercial feasibility. From outside Japan, weary looks were cast on those projects, suspecting that Japan, Inc. was aiming at global leadership in key future technologies, such as artificial intelligence and supercomputing. The reality was different. When there is no urgent need to develop a social and industrial infrastructure, the government-sponsored project tends to become a hobby club of scientists. Emotional arguments about the Japanese threat that were heightened in the U.S. amidst the flood of Japanese DRAM chips in the late 1970s through 1980s proved mostly groundless, as witness the outcome of the successors of the DEX and VLSI projects.

A mechanism or organization that works well in one phase of infrastructure development may well become ineffective in a different environment. Environmental change is precipitated by the working mechanism of the previous phase. The advent of certain technology is fostered by a certain environment; the growth of that technology changes the industrial infrastructure, which becomes the bed of next-generation technology. Unless we view technology development from the viewpoint of dynamics, we miss opportunities to resolve today's difficulties and build a harmonious world.

The dynamics of Japanese industry, however, has enormous inertia. In spite of a changing market environment, one aspect of the Japanese industrial machinery remains intact, while a schism between the domestic social mechanism and the international market mechanism is widening. If any significant restructuring ever occurs, it may have to be precipitated through wide and deep readjustments in culture and societal organization. The Japanese hierarchical industrial machinery proved efficient in catching up with the technology developments made mostly in the U.S. throughout the 1960s and 1970s. In a social environment where the digestion of imported technology was given a high priority, efforts tended to focus on improving the product yield. What mattered to people, from factory management down to the assembly-line worker, was improving the product quality. High-quality manufacturing and the resultant high reliability of products became the key to the competitiveness of Japanese products in the global market, almost regardless of the currency exchange situation. Due to the development of self-sufficiency in chip making, and also to the tendency to build excess capacity in manufacturing, the domestic market remained extremely tough for foreign competitors. Many critics from other countries regarded lopsided international trade in goods as a serious threat to the world's free-trade system.

The demand for change precipitated by Japanese manufacturing practices was non-technological. How can Japan increase imports? Political pressure and governmental arrangements corrected the trade imbalance to a certain extent, but did not provide any fundamental solutions. In approaching this trade issue, little effort has been expended to shed light on the technological perspective. It is natural for everyone to assume that the trade issue is in the realm of economists, government bureaucrats, and politicians. However, since no highly effective measures for correcting the imbalance have yet been devised, it may be worth considering the issue in the light of technological developments now in progress on the global scale. Would any technological reorientation or diversification on the part of Japan serve to open the domestic market to import? If so, it would affect the social organization of Japan, Inc. fundamentally.

Chapter 4

The Japanese Electronics Industry Today

The Electronics Industry Association of Japan forecasts the worldwide demand for audio-visual products through the year 2002. As shown in Table 4-1, demand for color televisions is projected to rise slightly, at 3.6 percent, according to replacement demand in the U.S., depending on the growth of digital broadcasting and high-definition television. The greatest potential for color TV growth is in the economic and income development of Eastern and Central Europe, the Middle East, Africa, and Asia. VTR growth is projected at 1.2 percent over the next five years, as the industry is limited to replacement sales. Primary growth is from digital VTR demand that will increase with digital broadcasting, and in developing areas like Central and Eastern Europe, Latin America, and Africa. Video CDs in China and cable television in India are impeding VTR growth.

Video camera demand is projected to grow at 5.3 percent through 2002. As prices continue to fall for new digital video cameras, demand is forecast to rise in Japan, the U.S., and Europe. Digital video cameras are projected to overtake conventional models by the year 2000. By the year 2002, digital video cameras are expected to account for approximately 75 percent of global video camera demand.

Digital versatile disk (DVD) players were expected to displace conventional video disk players by 2002, with growth rates averaging 70.7

Table 4-1 Global Audio-Visual Demand through the Year 2002 (millions of units).

Year	Color TVs	VTRs	Video cameras	Video disk players	Home audio	Portable audio	Car audio
1997	116.7	50.2	9.8	1.9	130.1	64.6	70
1998	122.2	51.1	10.4	3.1	131.7	65.6	71.1
1999	125.7	51.8	11.0	5.3	134.4	66.9	72.4
2000	130.4	52.1	11.5	8.3	137.2	68.4	73.4
2001	134.7	52.4	12.1	10.0	139.6	70.6	74.3
2002	139.3	53.4	12.7	11.5	141.7	73.2	75.1
Average growth rate	3.6%	1.2%	5.3%	43.0%	1.7%	2.5%	1.4%

Source: *EIAJ Review*, March 1998, p. 7.

55

percent annually. In the new entertainment market, DVD demand was expected to grow from 800,000 units in 1997 to 11.5 million units annually in 2002. Forecast growth was dependent on the major film production companies using DVD technology and the development of the DVD software rental business.

Home audio equipment growth was dependent on the addition of Mini Disk (MD) to stereo models for industrialized countries. Demand in Japan, the USA, and Western Europe has been leveling off for older technology cassette recorders with radios. Forecasted growth of 1.7 percent depended on growth in lesser developed countries in Asia, Latin America, and Central and Eastern Europe. Portable audio equipment is forecast to grow an average of 2.6 percent through 2002. With the maturing of home audio equipment, latent demand for portable equipment was expected in the U.S. and Western Europe. Portable headphone-type MD players were expected to increase market share against stereo cassette players.

Auto sales suffered with the Asian financial crisis, but recovery was expected in 1999, with average growth of 2.0 percent per year. Demand for CD and MD car players is projected to increase in the U.S., Western Europe, and Japan. Markets in Asia, Latin America, and Central and Eastern Europe are expected to grow rapidly as new car sales increase. Car navigation systems are projected to reach 3 million units in Japan, 1.5 million units in the U.S., and 2 million units in western Europe by 2000.

4.1. Maturity of the Japanese Electronics Market

Japan became a major player in the global electronics, electrical equipment, and components industries during the expansion of its consumer electronics, automotive, telecommunications, computer, and office automation equipment industries during the 1980s. As shown in Table 4-2, the penetration rate of electric and electronic appliances advanced in 1997 in all categories except stereos, which experienced a slight decline. PC ownership by Japanese consumers continued to grow. Further, use of facsimile machines in

Table 4-2 Ownership of Selected Products by Percentage of Japanese Households (1993-1997)

Electronics Products	1993	1994	1995	1996	1997
Color TVs	99.1	99.0	98.9	99.1	99.2
DBS (Direct Broadcasting System) tuner units	21.3	26.6	27.6	30.1	32.6
Video tape recorders (VTRs)	75.1	72.5	73.7	73.8	75.7
Compact disk players	54.3	53.8	55.9	56.8	57.9
Video cameras	25.6	29.9	31.3	32.3	33.6
Stereos	61.3	60.1	57.7	58.2	56.3
Japanese word processors	36.2	37.8	39.4	40.9	41.6
Personal computers (PCs)	11.9	13.9	15.6	17.3	22.1
Facsimile machines	6.7	7.6	10.0	12.9	17.5

Japanese homes was increasing, due to the introduction of models with a variety of new functions.

4.2. Japan's Electronics Market Position

With an estimated $223 billion in electronics sales, the Japanese market accounted for about 28 percent of the world market in 1998, compared to 48 percent for the U.S.'s market size of $375 billion. The exhibits in Tables 4-3 and 4-4 show Japan's market share by major electronics category. Japan represents 30 percent of the world market in electronic data processing, 25 percent of office equipment, 17 percent of controls and instruments, 26 percent of medical and industrial equipment, 23 percent of communications and radar, 33 percent of telecommunications, 27 percent of consumer products, and 29 percent of the world component business.

After Japan's consumption tax was increased in 1998, demand for household electrical appliances slowed. As a result of the economic slowdown, dealer inventories reached 2.5 million units in 1998 in Japan. Consumer spending had remained stagnant through mid-1998, with a record eight consecutive months of year-on-year declines in average household spending. With anxiety over the nation's financial crisis, the average propensity to spend - the ratio of consumer expenditures to disposable income - fell from 72.9 percent in April 1998 to 68 percent in March 1998. To reduce inventories, companies were suspending production and postponing new product introductions. As shown in Table 4-3, sale of VCRs and portable audio equipment actually declined by 1.1 and 0.4 percent, respectively, in 1998. There was moderate growth in television sets (1.1 percent), video cameras (4.3 percent), and car audio (2.1 percent). The only products to sustain prior growth

Table 4-3 Japanese Production of Consumer Electronics (000 units)

Consumer products (000 units)	1997 sales	% change from 1996	1998 production forecast	% change from 1997
Televisions	11,500	0.8	11.630	1.1
VCR	7,600	3.2	7,520	-1.1
Video cameras	1,380	4.9	1,440	4.3
VD player	249	7.8	560	124.9
Home audio	6,699	-1.6	6,705	0.1
MD stereo	314	33.1	350	11.5
CD stereo	457	24.2	200	-56.2
Portable audio	7,368	-0.7	7,340	-0.4
Car audio	10,090	4.9	10,300	2.1
Car navigation	900	15.4	1,200	33.3
Car televisions	779	-2.9	850	10.4

Source: *NIPPON 1998: Business facts & figures*, JETRO, June 8, 1998, p. 34

rates of over 10 percent were non-portable MD players (11.5 percent), car televisions (10.4 percent), car navigation systems (33.3 percent), and new video disk players (124.9 percent).

Increased introduction of multimedia and new digital devices, including the PHS and personal digital assistant (PDA) has sharply increased the demand for "system LSIs" in which high-efficiency, multi-functional devices, including smaller, low power-consumption types and high-speed data communications equipment, are integrated into a single chip. With the traditional market for DRAMs stagnating after the 1996 price crash, Japanese semiconductor makers moved into the system LSI arena, considered "the New Industrial Revolution" in electronics. Manufacturers were planning to build a mixed production system in which system LSIs and DRAMs could be manufactured on the same production line.

4.3. The Micro-electronics Industry Structure

The global semiconductor industry has seen a shift in competition to developing countries in the Asia-Pacific area outside of Japan and North America. The goal for semiconductor production in Taiwan and Singapore is 15 percent of the global semiconductor industry's market share, 7.5 percent each, by the early part of the twenty-first century. As shown in Table 4-4, market share had already doubled, from 6.0 percent to 12.1 percent, between 1992 and 1995. Due to increased competition, industry prices fell sharply and profitability suffered. While overall growth continued, the new Asian competitors were taking an increased share of the growth.

As shown in Table 4-5, five of the top ten semiconductor firms were Japanese firms: including NEC, Toshiba, Hitachi, Fujitsu, and Mitsubishi; three were U.S. firms (Motorola, Texas Instruments, and Intel); one was South Korean (Samsung); and one was European (Philips). The leading semiconductor firm was Intel, which dominated the microprocessor business.

Table 4-4 Global Market Shares for Semiconductors (Unit: US$ billion)

Year	1985	1986	1987	1988	1989	1990	1991	1992	1993	1994	1995
Share of Japanese Companies	101	142	185	259	298	269	277	276	346	448	611
	42%	46%	48%	51%	52%	46%	46%	42.3%	40.4%	40.5%	39.5%
Share of North American Companies	111	128	149	186	200	225	229	271	372	461	616
	45%	42%	39%	37%	35%	39%	38%	41.5%	43.4%	41.7%	39.8%
Share of European Companies	29	34	42	49	54	65	63	67	77	98	133
	12%	11%	11%	10%	10%	11%	11%	10.2%	9.0%	8.9%	8.6%
Share of Asia Pacific Companies				14	20	23	28	39	62	98	186
	31%	42%	72%	3%	4%	4%	5%	6.0%	7.2%	8.9%	12.1%
Total	243	308	383	509	572	582	597	653	856	1,106	1,547

Source: Data Quest

Intel increased its 1997 lead to $21 billion in the $150 billion worldwide chip market. The consolidated earnings of Japan's electronics leaders declined in 1997 and were relatively flat for 1998, though profitability improved in 1998. As shown in Table 4-6, NEC was investing 300 billion Yen in facilities to produce 256-megabit DRAM chips in Japan and the U.S. by the year 2000.

Nearly two-thirds of NEC's semiconductor business is in LSI chips used in popular products like mobile phones and car navigation systems. Fujitsu was withdrawing from the DRAM business, increasing flash memory and LSI chip production. Hitachi expected group net profit to surge 1,050 percent for 1998 as its consumer product business improved and semiconductor losses declined. Toshiba had suffered for declines in its power-related business, especially in Southwest Asia, but forecasts higher profits as sales in notebook PCs and appliances improved. Matsushita, relying on consumer electronics, expected

Table 4-5 Leading Semiconductor Firms

'96 Rank	'97 Rank	Company	'96 Revenue	'97 Revenue	%Change
1	1	Intel	$17,781	$21,083	18.6%
2	2	NEC	$10,428	$10,656	2.2%
3	3	Motorola	$8,076	$8,120	0.5%
6	4	Texas Instruments	$7,064	$7,660	8.4%
5	5	Toshiba	$8,065	$7,507	-6.9%
4	6	Hitachi	$8,071	$6,523	-19.2%
7	7	Samsung	$6,464	$6,010	-7.0%
8	8	Fujitsu	$4,427	$4,872	10.1%
9	9	Philips	$4,219	$4,435	5.1%
10	10	Mitsubishi	$4,100	$4,097	-0.1%

Table 4-6 Electronics Firm Results for 1998

Electronics Companies	1998 Sales	% change	Net Profit/ Loss	% Change
NEC Corp.	5,100.0	4.0	55.0	33.0
Fujitsu Ltd.	5,600.0	12.0	75.0	1,263
Hitachi Ltd.	8,400.0	0.0	40.0	1,050
Toshiba Corp.	5,550.0	2.0	40.0	445.0
Mitsubishi Electric Corp.	3,950.0	3.0	20.0	--
Matsushita Electric Ind. Corp.	8,000.0	1.0	128.0	37.0
Sony Corp.	6,850.0	1.0	215.0	-3.0

Source: Kazunari Yokota, *Electronics Forecasts Challenged, The Nikkei Weekly*, June 1, 1998, p.6

video and audio equipment sales to increase. Industry analysts were less optimistic about their forecasted performance. Indeed, the current situation is very fluid; e.g., Hitachi announced in September 1998 an expected loss of 100 billion Yen for 1998.

4.4. Miniaturization of Electronic Products

In Japan, unlike the US, most companies that are heavily invested in electronics product markets are also vertically integrated and heavily invested in upstream semiconductor technologies. Six vertically integrated firms produced 85 percent of Japan's semiconductors, 80 percent of Japan's computers, 80 percent of Japan's telecommunications equipment, and 60 percent of Japan's consumer electronics products. Figure 4.1 shows the value contribution across the electronics value chain in 1990 for the US electronic industry [Gover and Gwyn, 1992]. US firms have been more successful in component technologies than in end product markets. While upstream electronics technology has a significant impact on downstream markets development, it is the product demand that drives technology development.

In response to demands for miniaturization and for thinner and lighter consumer electronic products, vendors of electronic devices and parts have been directed to reduce size or increase the number of components per square centimeter while increasing the number of electronic functions available. Miniaturization has been accomplished through the reduction in part sizes, including reductions in battery and display dimensions that allow for designs of lighter and smaller product profiles. For example, the size of cellular phones was reduced from 500 cubic centimeters in 1987 to 150 cubic centimeters in 1991. By 1998, cellular phones were under 70 cubic centimeters. Reductions came from a variety of improvements including:

- Reduced power amp size by reducing part sizes and ICs
- Reduced filter size by reducing part sizes and using advanced designs
- Reduced controller size through special ASIC development
- Reduced battery size by reduced power consumption requirements and improved battery life
- Reduced circuit board size by using smaller and thinner pin pitch LSIs, use of ball grid array packages, and smaller, more integrated passive parts

Other components are being developed to facilitate miniaturization. In mid-1998, Matsushita introduced a multifunctional cellular phone and pocket-pager speaker that includes vibration, loudspeaker and ringer in a

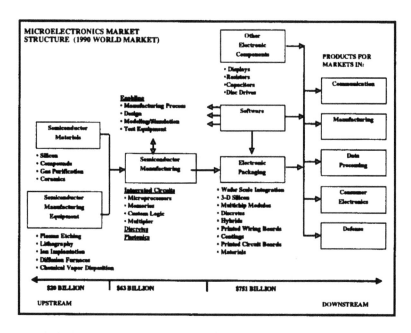

Figure 4.1 Value contribution across the electronics value chain in 1990 for the US electronic industry [Gower and Gwyn, 1992]

single 3.8-gram unit. The device requires half the space and one-third the number of assembly steps of alternative components.

Miniaturization of consumer products has driven developments in electronic parts and packaging technologies. Control of electronic packaging technologies provides Japanese companies with the competitive capabilities to design and manufacture smaller and more sophisticated consumer products, like camcorders, cameras and digital data books. With the innovative merging of semiconductor, packaging and display technologies, we are also seeing the evolution of second and third generation technologies that will increasingly affect the design parameters of future products. Those firms that have capabilities in materials, equipment, design and advanced manufacturing will be the future producers of low-cost electronic packages. SONY has been the world leader in miniaturization and has provided the miniaturization roadmap for Japan's component and equipment makers for nearly 40 years.

A firm's involvement in downstream products and markets provides the vision for setting priorities for developing future upstream technologies. Electronic packaging and other component technologies are becoming the critical technologies for advanced product designs and functions. Firms that dominate large market segments have been investing heavily in the technologies that will determine the design parameters for future product and system applications. For example, Japanese firms have developed dominant positions in display technologies, in the miniaturization of low cost, high-volume electronic packages, and in the development of advanced manufacturing equipment for ultra-small component assembly. In order to introduce its "next generation" products, Japanese firms have had to push their development of:

- Fine-pitch devices
- Advanced flex circuit designs
- Cost effective direct chip attachment technology
- Cost effective, high density printed circuit boards
- Cost effective connector technology
- Advanced flat panel display technology
- Miniaturization and integration of passive components
- Advanced, low cost package assembly equipment
- High-speed pick and place equipment for ultra-small components
- Improved battery technology

These developments have provided Japanese firms with the ability to both design-in greater functionality and performance for any given size product, or to reduce the size or cost of products at any given level of functionality or performance. These "intermediate" packaging capabilities provide much of the value added for video cameras. In fact, SONY Corporation found that 65 percent of an electronic product's value added derives from key component and device contributions as compared to only 12 percent coming from final assembly. Thus, SONY saw the advantage of producing more key components as a strategy to improve profitability. To focus on component development, firms also needed a roadmap of future product concepts or next generation product developments.

SONY Corporation's historical success in audio-visual markets was based on the introduction of the transistor radio in 1955, the Trinitron in 1968, the Betamax in 1975, the Walkman in 1979, the CD player in 1982, and the 8-mm camcorder in 1985. SONY's introduction of the Trinitron color television was dependent upon the development of its single electron gun with three electron beams and the aperture grill. After 8 years of development, SONY's technological breakthrough was considered the next generation beyond the shadow mask developed by RCA. The Trinitron technology continues to provide SONY with a basic technology for use in advanced high definition television applications. SONY also developed a charge couple device (CCD) for image sensing. After two years of basic research, SONY took five years to produce the first small CCD video camera for use in the industrial market. It took five more years to develop the mass production technology required to enter the consumer market. The resulting miniaturization of the video camera gave SONY a dominant position in the 8-mm video camera market.

Sharp Corporation has used a similar key device development strategy. Sharp introduced the first commercial application of a liquid crystal display (LCD) in an electronic calculator in 1973. Sharp pioneered the application of LCDs in Japanese word processors and PCs in 1986, developed the world's first 14" color TFT LCD panel in 1988, and developed the LCD video projector in 1989. Sharp's Hi-8 LCD ViewCam, introduced in 1992, was the world's first camcorder with a 4" color TFT LCD monitor. In 1993, Sharp's ViewCam surpassed both Matsushita and JVC to take second place in the video camera market. SONY responded by adding LCDs to its traditional camcorders. The ability to take leadership in critical component technologies is essential in developing market leadership.

With their early success in consumer electronics products, Japanese firms created a tremendous demand for upstream electronic technologies including DRAM memory chips and many intermediate electronic component and parts technologies. Growing domination of these intermediate component and packaging technologies, including liquid crystal display technology, provided strong business opportunities as final assembly moved off shore. Advanced display technology, the integration of both display and packaging technologies, is expected to generate opportunities for new consumer products ranging from hand-held computers to high definition television sets.

In looking at future technologies, six areas have been identified for development, including opto-electronics, electronics packaging, electronics and semiconductor materials, display technology, ULSI fabrication equipment, and memory chips. It is estimated that these six technologies will represent a world market worth over $100 billion by the year 2000. What is more important, it is expected that whoever dominates these areas of technology will also dominate the world's electronics markets that are expected to exceed $2 trillion in revenues.

Even more critical is the fact that Japanese companies do not intend to give their competitive advantages away. Most recently, SONY has limited visits by outsiders to its operations for fear of losing proprietary information. In cases where outsiders are allowed to visit SONY, they are shown production technology that the company is selling to outside customers. As with Matsushita, the world leader in semiconductor insertion equipment, advanced equipment development provides the company with multiple years of use before being made available to the open market. As product life cycles continue to shorten, six months or one-year delays in product availability can mean the difference between substantial profits and significant losses over the short life of the product. Unfortunately, Japanese firms have had difficulty taking advantage of their own technological leadership. Heavy regulation and government intervention in Japan's business activities has shifted the advantage in many electronics products to more responsive countries like Taiwan, or lower cost countries like China.

In the future, pragmatic business executives are not likely to sell their advanced micro-electronics technologies or advanced equipment if it creates competitors in critical downstream markets. As Texas Instrument's deceased chairman, Patrick Haggarty, reflected, "TI's biggest mistake may have been in not integrating forward into portable radios in the early 1950s." It was that same development of portable radios that gave SONY its first major success in the consumer products market.

4.4.1. Electronic Packaging Technologies

Electronic packaging includes all of the electronic components, assembly equipment and processes required for the production of electronic products. It includes the traditional packaging of semiconductor devices in both single and multichip configurations, all forms of passive components, the various types of flexible and rigid circuit boards used for mounting components, and the variety of techniques and equipment used for overall electronic package assembly and testing operations. The primary concern for electronic packaging configuration has to do with the ultimate trade-offs between the function, performance, design and cost that are required in the final product.

Electronic packaging technologies can be described in terms of materials, processes, components, subsystems and end products or systems. Miniaturization of product designs has pushed technologies to deal with the physics and functional density requirements of miniaturization, causing new materials, components and processes to be developed in response to ultra-small and ultra-lightweight components. At the materials level, Japan leads in both epoxy and ceramic materials development used in the packaging of semiconductor packages. With the move to fine-pitch lead packages, solder pastes have been developed using fine spherical-grain metal powders rather than irregular-grain powders. To produce miniaturized active and passive components, materials technologies have been pushed to their limits.

In parallel with materials developments, production and assembly equipment have been developed for component production and packaging assembly. Miniaturization of consumer products has made Japan the leader in precision assembly technologies. In molding systems, multiplunger systems are currently being used as new systems are being proposed, such as "3P systems" of BDM, "Cullless system" of Tamus Tech., and the "Vacuum mold systems" of Towa. "Clean pellet" molding materials offer less volatility and better moldability. Assembly of miniature components makes soldering performance the number one issue, with a move from wave soldering to reflow soldering methods. New low cost, low stress soldering compounds are essential for continued miniaturization.

At the components level, requirements for miniaturization, power saving, improvement of reliability, and cost reduction are today's key issues. Miniaturization through circuit integration improves energy and performance characteristics of most electronic products. Sony's new cellular phone, the size of a credit card, utilizes a one-chip solution for ROM, RAM, CPU, and ADC. Portability has also required Japanese firms to develop lightweight, low-energy consuming components like LCDs. Japan is the world leader in LCD production for lightweight, low-energy consuming, portable consumer products. Such products put pressure on reduction of the space used for electronic components, leading to further developments in low profile or thin quad-flat-packages, tape-carrier packages, multilayer passive components, and small-sized, high energy density lithium ion batteries for consumer electronics.

At the product level, Japan continues to lead the world in advanced consumer electronics products. The U.S. continues to lead Japan in communication and information technologies. Japan is pushing to catch up in these areas. This notwithstanding, Japan leads in most multimedia products with the merging of consumer products into miniaturized combination products. For example, in 1994, Sharp Corporation introduced infrared facsimile communication for its Zaurus (Sharp's version of the Apple Newton personal digital assistant). The simple integration of multi-media technologies into consumer products defines Japan's future product strategy as AVCCC -- the combination of audio, visual, computers, communication, and controllers. The integration of these functions will require further miniaturization of final products for improved portability.

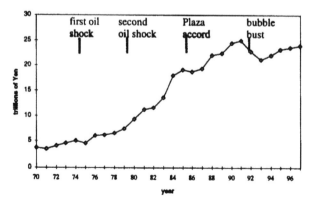

Figure 4.2 Histogram showing the total production by the Japanese electronics industry from 1970 to 1997 [Nikkei Electronics, 1996]

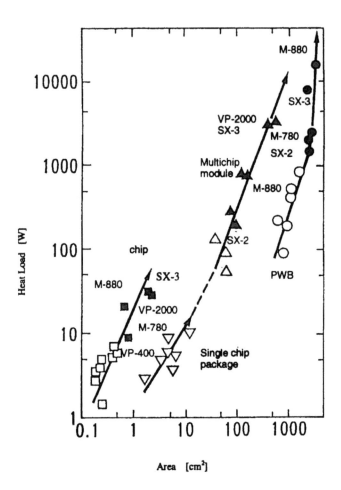

Figure 4.3 Trend toward increasing heat dissipation from LSI logic chips, modules, and priting wiring boards (PWB) for mainframes and supercomputers [Nakayama, 1992]

4.5. Overall Landscape

In this section the state of the art of the Japanese electronics industry is reviewed in light of the technological and market developments in progress worldwide. Distinct developments of the 1990s include accelerated dominance of microprocessor-based technology, the advent of a global communications infrastructure, the concomitant rise of the software service industry, and the emergence of other Asian countries as Japan's competitors.

Section 4.1 depicts the overall landscape of the industry today. Section 4.2 illustrates the mechanisms for product design and development currently at work in the Japanese electronics industry, using an example from the design of portable computers. Section 4.3 discusses the relationship between technology and industrial organization, an issue mentioned in the final paragraph of Chapter 3.

Figure 4.2 is a histogram showing the total production by the Japanese electronics industry from 1970 to 1997 [*Nikkei Electronics No. 675*, 1996]. The first oil embargo by Arab nations in 1974 left a dent in the growth curve; however, the oil price hike of 1979 had little effect on growth. From the early to mid-1980s, rapid growth was driven chiefly by exports. The Plaza Accord among the G7 nations in 1985 invited a rise in the Yen against other currencies, thereby setting a brake on exports, but it produced only a respite. A true turning point came when the economic boom driven by land speculation, called the bubble economy, burst in the early 1990s. This coincided with the movement toward downsizing computers.

4.5.1. Semiconductors

The price of mainstream memory chips (4M-bit and 16M-bit DRAM chips) dropped by one-fifth in 1996, [*Nikkei Electronics No. 675*, 1996] then stayed in the low-level range in 1997.

Since 1996, growth in the world's semiconductor market has been sluggish since 1996 due to reduced DRAM demand. The market in Japan has been no exception, down 10.3% in 1998 compared with 1997. Market value has fallen to $34.159 billion (4.1268 trillion Yen, a 0.5% increase based on the Yen).

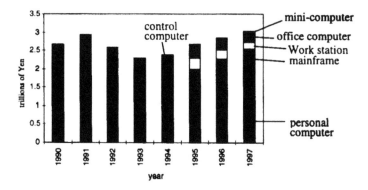

Figure 4.4 Histogram showing personal computers (PCs) dominance in the latter 1990s [Nikkei Electronics, 1996]

The major culprit for this sluggish growth is the DRAM market, which has fallen into a significant negative growth of 22.2% compared to the previous year (based on the Yen). Memory products other than DRAMs have also recorded negative growth, further aggravating the sluggish growth of the Japanese market. The negative growth for mask ROM and EPROM, for example, have been significant, causing overall growth of MOS memory products to register a 15.8% drop. The MPU has also fallen into negative growth of 9.4% compared with the previous year, from the high growth rate of 38.8% in the previous year, due mainly to sluggish growth in the PC market and a drop in per unit prices in the Japanese market.

There is no prospect of a hike in the per-bit price, in spite of the forthcoming switchover of leading DRAM chips from 16M-bit to 64M-bit. The glut of DRAM chips has resulted from the expansion of manufacturing capacity in Japan, Korea, and Taiwan. From 1994 to 1995 the shipment of DRAM chips from Asia overall increased by sixty percent. Facing diminishing returns, Japanese chip-makers began slowing down domestic capital investment in new factories and equipment. In terms of the percentage reduction in investment compared to that of the previous year, common figures across the Japanese chip makers are about eight percent in 1996, and four percent in 1997 [*Nikkei Electronics No. 702*, 1997]. Hitachi, Fujitsu, and Mitsubishi Electric began entrusting DRAM chip production to Taiwanese foundries, such as Taiwan Semiconductor Manufacturing Company, Ltd. (TSMC) and United Microelectronics Corporation (UMC), independent multi-client chip makers that sprang up in 1996 and 1997. While Taiwan chip makers are gaining experience through OEM production, the Taiwanese government has been encouraging the development of design capability [Lee and Pecht, 1997] in such companies as Mosel Vitelic, Vanguard, and Taiwan Memory [*Nikkei Electronics No. 702*, 1997]. NanYa Technology Corporation, Powerchip Semiconductor Corporation, and Texas Instruments-Acer Inc. (this joint venture dissolved in 1998) are also in the forefront of chip manufacturing.

Korean companies have been adopting very aggressive expansion plans. From 1996 to 1998, the pace of new plant construction by three DRAM makers (Samsung, LG Semiconductors, and Hyundai Electronics) has been one per year. The construction of all those plants is self-financed, except for one Samsung plant in Texas that is partly financed by Intel. The technology of these companies is also becoming self-reliant, covering all steps from design, processing, and packaging to inspection. LG Semiconductors, for instance, is planning to reduce its OEM supply to Hitachi as it switches to 64M-bit production [*Nikkei Electronics No. 702*, 1997].

Table 4-7 About 70% of domestic produced DRAM chips are exported [Nikkei Electronics, 1996]

	Production (P) (trillion Yen)	Annual growth (%)	Export (E) (trillion Yen)	E/P (%)	Import (I) (trillion Yen)	Domestic consumption (D) (trillion Yen)	I/D (%)
1984	2.584	165.6	0.897	34.7	0.270	1.957	13.8
1985	2.410	93.2	0.697	28.9	0.199	1.913	10.4
1986	2.332	96.8	0.639	27.4	0.173	1.866	9.3
1987	2.487	106.6	0.724	29.1	0.189	1.952	9.7
1988	3.119	125.4	1.015	32.5	0.257	2.361	10.9
1989	3.594	115.2	1.337	37.2	0.352	2.609	13.5
1990	3.623	100.8	1.306	36.0	0.424	2.741	15.5
1991	3.887	107.3	1.342	34.5	0.467	3.012	15.5
1992	3.419	88.0	1.488	43.5	0.442	2.373	18.6
1993	3.550	103.8	1.706	48.1	0.522	2.366	22.1
1994	4.037	113.7	2.167	53.7	0.676	2.546	26.5
1995	4.793	118.7	2.813	58.7	1.082	3.061	35.3
1996*	4.585	95.7	3.197	69.7	1.347	2.734	49.3
1997**	4.756	103.7	3.334	70.1	1.595	3.018	52.9

*   Including estimates by Nikkei Electronics
** Estimates by Nikkei Electronics

The ratio of the Japanese market to the world market has also fallen sharply from 25.9% in 1996 to 22.7% (based on the U.S. dollar) in 1997. The ratio of the Japanese market had been on a downward trend for the past several years due to the rapid growth seen in other Asian markets. In 1997, the downward trend grew significantly worse.

The rise of Asian chip makers, and also the strengthening position of Texas Instruments and Micron Technology, Inc., have contributed to the slippage of Japanese chip makers in the rankings for DRAM sales. In the sales rankings for the second quarter of 1997, Texas Instruments was well ahead of other competitors, followed by Samsung, Micron Technology, NEC, and Hyundai [*Nikkei Electronics No. 702*, 1997].

The relative decline of Japanese firms in the rankings, however, does not alter the characteristics of the industry in any fundamental way. As Table 4-7 [*Nikkei Electronics No. 663*, 1996] shows, around seventy percent of the domestic production of DRAM chips is being exported. The volume of imported chips is also increasing, due primarily to the shift of production of older generation chips to overseas plants. Table 4-8 [White Paper of Trade and Commerce, 1996] shows that DRAM chips have captured the largest percentage of the domestic production volume of chips, accounting for some eighteen percent as of 1994.

Table 4-8 Semiconductor production by Japanese companies [White Paper of Trade and Commerce, 1996]

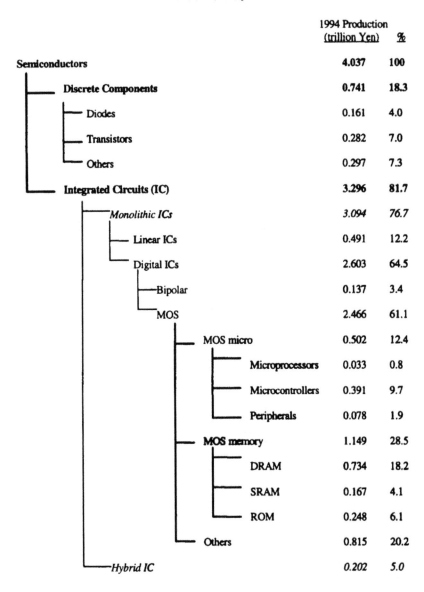

	1994 Production (trillion Yen)	%
Semiconductors	4.037	100
Discrete Components	0.741	18.3
Diodes	0.161	4.0
Transistors	0.282	7.0
Others	0.297	7.3
Integrated Circuits (IC)	3.296	81.7
Monolithic ICs	3.094	76.7
Linear ICs	0.491	12.2
Digital ICs	2.603	64.5
Bipolar	0.137	3.4
MOS	2.466	61.1
MOS micro	0.502	12.4
Microprocessors	0.033	0.8
Microcontrollers	0.391	9.7
Peripherals	0.078	1.9
MOS memory	1.149	28.5
DRAM	0.734	18.2
SRAM	0.167	4.1
ROM	0.248	6.1
Others	0.815	20.2
Hybrid IC	0.202	5.0

The item with the second largest percentage in Table 4-8 (9.7%) is the microcontroller. Japanese chip makers are strengthening their development and production of microcontrollers for diversified applications [*Integrated Circuit International*, 1997] such as multimedia processing, computer peripherals, automobiles, camcorders, and mobile telephones. NEC, Mitsubishi Electric, Hitachi, Toshiba, and Fujitsu all have plans to double their design and production capability between 1997 and 2000.

The rate of investment for the information industry has generally slowed due to the worsening profit outlook of the Japanese industry. Sales of consumer electronics equipment have fallen due to the decrease in consumer purchases caused by the rise in consumption and cut-back or minimal increase in payroll due to the poor profit outlook of the industry. In other words, the Japanese economy has fallen into a severe vicious cycle, which is directly affecting the electronics equipment market.

The Japanese market is forecasted to recover significantly after 1999. However, the share of the Japanese market in the world market is expected to decrease steadily, since its growth rate is expected to grow at a lower rate than in other regions. It was forecasted that it would drop steadily from 22.7% in 1997 to 18.0 % in 2002, resulting in the lowest rate compared to the other three major regions (North America, Europe and Asia).

4.5.2 Semiconductor Equipment

The Japanese semiconductor equipment market in 1997 was estimated to be $6.99 billion, a 5.1% decrease from the previous year (838.46 billion Yen, a 4.7% increase based on the Yen). Manufacturers of mainly DRAM products held off on capital investments due to the soft DRAM market. Consequently, a small negative growth from the previous year was recorded. The overall world market is forecasted to be $24.38 billion, a 5.2% increase over the previous year to 28.7%. Japan had been the world's largest regional market, but the top position will now be held by the North American market, constituting 31.0% of the world's market.

Looking at the overall market by type of equipment, front-end equipment such as thermal processing equipment, ion implantation equipment, CVD equipment, etc., are experiencing negative growth or minimal growth in 1998. Thermal processing equipment (oxidation, diffusion furnaces, and RTP equipment) showed a market value of 31.7 billion Yen ($U.S. 262 million), a 12.2% decrease. The ion injection equipment showed a market value of 27.6 billion Yen ($U.S. 228.5 million), a 4.8% decrease, and CVD equipment showed a market value of 86.8 billion Yen ($U.S. 718.6 million), a 9.1% decrease, all exacerbating negative growth of the overall market for manufacturing equipment.

On the other hand, CMP equipment recorded a high growth rate of 72.4%. Additionally, assembling and testing equipment also recorded double-digit growth. Assembling equipment showed a market value of 58.7 billion Yen ($U.S. 485.9 million), an 18.6% increase, and testing equipment showed a market value of 177.7 billion Yen ($U.S. 1.47 billion), and 11.1% increase.

The Semiconductor Manufacturing Equipment Association of Japan (SEAJ) has announced its forecast results showing negative growth for the 1998 manufacturing equipment market compared with 1997. The value of the Japanese manufacturing equipment sales (total value of domestic and overseas sales by Japanese companies including those operation overseas) is estimated to be 4.48 trillion Yen ($U.S. 9.35 billion), or 85.4% of the previous year value. Total Japanese market sales value (combined total of Japanese companies' manufacturing equipment aimed at the domestic market and foreign capital companies manufacturing equipment aimed at the Japanese domestic market) is estimated to be 630.7 billion Yen ($U.S. 5.22 billion) or 84.7% of the previous year value.

Those who plan to introduce 300 mm wafers are divided into aggressive and cautious support groups. Here are the construction plans and future prospects revealed by key Japanese companies:

- NEC will construct a new plant to produce 300 mm wafers on the present site of NEC electronics Roseville Plant (California), NEC's U.S. subsidiary. The investment value will be $1.4 billion (180 billion Yen). This will be the production base for system-on-chips and 0.15 μm devices such as the 1G DRAM. It is scheduled to commence operation in 2002, and about 700 new jobs will be created. The production capacity is expected to be 20,000 wafers per month, and the floor space will be about 600,000 ft^2.
- Toshiba had not disclosed any specific plans, but there is strong indication that mass production will start in 2001 to 2002. Koichi Suzuki, manager of the Semiconductor Business Division and director of Toshiba, said at the executive symposium, Semicon Kansai 98, held in June, "we are not denying that plans are in place to produce 300 mm wafers, but it should be made clear that the refinement of pixel size should be given first priority."
- Hitachi claims it will begin construction of new manufacturing lines in 1999 while closely watching the market situation. The tentative base of production is the Naka plant (Ibaragi Prefecture). Advanced DRAM succeeding the 25M DRAM, system LSI, large capacity flash memory, SH microcomputer, etc., will be produced. The expected production capacity is about 10,000-20,000 wafers per month. The capital

investment in the facility is expected to be on the order of 100 billion to 150 billion Yen ($U.S. 827 million to $U.S. 1.24 billion).

- Fujitsu may postpone the start of operation at its new Aizu-Wakamatsu plant (Fukushima Prefecture) from 2001 to 2002. This plant includes a floor space of 18900 m^2 and a cleanroom area of 15,000 m^2. The initial production capacity will be 15,000 wafers per month with 0.18 μm design rules. The items produced are expected to be 256M and IG DRAMs and logic devices. The capital investment is expected to be on the order of $U.S. 1.2 billion

- Mitsubishi Electric reports that construction of a prototype production line will be in 2000 at the earliest. Initially it will aim at producing 0.18 μm products on the 200 mm wafer production line, then decide on construction of the 300 mm wafer production line, depending on the market situation. Like Toshiba, Mitsubishi also intends to place higher priority on refined pixel size rather than production of large diameter wafers. It reports the design rule of the 300 mm wafer production line could be 0.18 μm, or even startup based on a 0.15-0.13 μm rule. A new plant building will be constructed at the Kochi Plant (Kochi Prefecture) for mass production.

- Oki Electric Industry is preparing to launch a new production line after 2000, depending upon the market situation. It is in the process of acquiring a suitable site for its production base in Japan. Whether 300 or 200 mm wafers will be produced in the new production line is unknown.

- Matsushita Electronics industry is placing its priority on the study of a small-scale plant. The feasibility of producing 300 mm wafers will be investigated as an extension of this study.

- Sharp is investigating the feasibility of producing 300 mm wafers at its main production plant, Fukuyama Plant (Hiroshima Prefecture).

4.5.3. Computers

In the computer market, the traditional customers for mainframes had been banks and large securities trading companies. Purchase orders from such buyers evaporated with the bubble bust. Meanwhile, manufacturing mainframes became costly as the machines increasingly needed elaborate hardware, such as large multichip modules, multilayer wiring boards, and efficient cooling devices. For instance, Figure 4.3 shows the trend toward increasing heat dissipation from LSI logic chips, modules, and printed wiring boards (PWB) for mainframes and supercomputers. The horizontal axis shows the size expansion of those components. The data were collected from publications from Fujitsu, NEC, Hitachi, and IBM Japan during the period from 1970 to 1990. [Nakayama, 1992]. Open data points mean that the

CPUs of those computers were air-cooled, while solid points indicate elaborately designed water-cooled computers.

After 1990, these shooting trends abruptly stopped as manufacturers stopped pursuing this kind of technology. Large systems shipped in the 1990s were those designed in the 1980s. In the early 1990s, lease or purchase orders carried over from the 1980s kept the share of mainframes in the total shipment of computers relatively large, but since 1994 that share has shrunk at an ever greater rate (Figure 4.4), with a twenty percent decrease in 1996 and a thirty percent decrease in 1997. Large systems, however, will not disappear. CMOS-based machines are on an upward trend, with a six percent annual increase, as observed by NEC. Also, in Fujitsu, NEC, and Hitachi the mentality still remains of betting the company's pride on juggernauts. From time to time, announcements of the world's highest processing speed record are made through news media, with complimentary comments from well-known academics and bureaucrats. Distributed-memory parallel processing has now replaced brute-force high-speed processing on ECL gate arrays. IBM Japan and Cray Research also have sizable shares in high-performance computing.

Figure 4.4 [*Nikkei Electronics No. 675*, 1996] shows the obvious take-over of personal computers (PCs) as the computer market leader in the latter 1990s. In 1997, the production volume of PCs reached 8,163,000. So far, high demand for PCs has been generated by the corporate sector, particularly, large companies eager to improve the productivity of white-collar workers. Nevertheless, information technology represents only a modest share (10%) of the total Japanese technology investment, compared with its share in the U.S. (40%) [*Nikkei Electronics No. 704*, 1997]. Also contributing to the rise in the production volume of PCs is the back-flow of production from overseas plants to domestic ones. Ever-accelerating product development cycles, releasing about three new models a year, make domestic production a more economically sound option than overseas production. The most obvious benefit of this move is the ease of inventory control.

Sales of PCs, however, showed a sign of a sudden stall in the latter half of 1997 [*Nikkei Electronics No.704*, 1997]. This has been attributed to a failure to cultivate more home PC users. In terms of computer ownership, Japan lags behind most of the advanced nations, as of 1995, about 150 computers per thousand people in Japan, less than half the volume in the U.S. [*Nikkei Electronics No. 704*, 1997]. In the short term, the blame lies with the sales tax hike of April 1997 and the high cost of communication through telephone lines, which is about three times that in the U.S. In particular, the latter has been a constraint on the introduction of commercial Internet service to Japanese homes. The same culture clash that has shaped the history of Japan's electronics industry pertains to Internet development, which has been affected by the tug of war between the centralized bureaucratic system (which is least enthusiastic about deregulation) and small service providers. In spite of the odds, hundreds of Internet service providers have sprung up. The number of Internet subscribers is increasing rapidly, at a rate of about fifteen percent per month.   Many analysts expect this surge of Internet subscriptions to rev up the PC market in the late 1990s, if the overall economy does not go sour.

In contrast to the overall slowdown in PC sales, portable PCs are hot items.   Portable PCs account for a substantial share of Japan's PC market, estimated to be forty percent in 1998 [*Nikkei Electronics No. 704*, 1997].   In cramped business offices, portable PCs are favored over desktops.   Those designed for office environments are called "high-end" notebooks.   The mainstream features of these portables are 800 x 600 pixel TFT displays and microprocessors with clock frequencies reaching 200MHz.   Portable PCs for personal use are shrinking in size, weight, and power consumption.   In 1997, Toshiba released the "Libretto" model, the smallest ever in the history of portable PCs, which measures 21cm x 11.5cm x 3.4cm and weighs 850 grams.   Major competitors in the portable PC market are NEC, Fujitsu, Toshiba, Sony, IBM Japan, Sharp, Matushita Electric, and Compaq.   The intensity of price cutting competition in the portable PC market is comparable to that in other consumer electronics markets. This is also where miniaturization technology is applied in its most advanced forms, as we will see in Section 4.6.

Domestic production of workstations (UNIX-based computers) is also decreasing. Sun Microsystems and Hewlett-Packard are about to monopolize the workstation market through their Japanese subsidiaries, which are importing most of the workstations and applications software for engineering purposes. The function of workstations is being pushed to the higher ends — namely, three-dimensional graphics and UNIX servers for Internets and intranets — where PCs have not yet eroded the market.

Office computers are medium-size machines for business use.   The primary function of office computers has been shifting from stand-alone data processing to server use.   NEC's Express 5800 series with Windows NT became a hit in 1996.   High demand is expected from those medium-size firms where UNIX servers are regarded as too expensive, while PC servers fall short of the performance necessary to deal with the increasing number of client PCs.

4.5.4. Computer Peripherals

Figure 4.5 [*Nikkei Electronics No. 675*, 1996] shows a histogram of the domestic production of computer peripherals.   In terms of growth rate, displays and optical discs are two leading items. Other components have not shown any remarkable growth because of the shift of production to overseas

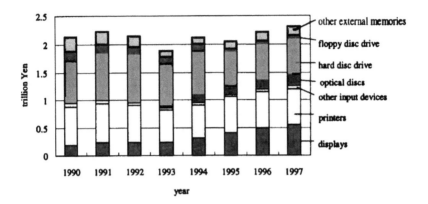

Figure 4.5 Histogram of the domestic production of computer peripherals [Nikkei Electronics]

plants. Figure 4.6 [*Nikkei Electronics No. 675*, 1996] shows the percentage of overseas production by Japanese makers for various consumer electronics items as of 1995. More than ninety percent of floppy disc drives (FDDs), forty-five percent of hard disc drives (HDDs), thirty-five percent of CD ROMs, and twenty-five to twenty-eight percent of ink-jet and optical printers were produced in overseas plants.

Liquid crystal displays (LCDs) have undergone cycles of glut and shortage, each phase extending about a year and a half. The latest period of shortage was from the second quarter of 1996 to the third quarter of 1997, when the switch was made to 12.1 inch displays on notebook PCs [*Nikkei Electronics No. 704*, 1997]. 1998 is a period of glut, as Japanese and Korean makers increase supplies, while sales of notebook PCs may not pick up proportionately. In the fourth quarter of 1997 the worldwide production of color TFT displays reached a million panels per month, an increase of 250,000 panels per month compared with the previous quarter. Almost all of the increase was produced by newly built NEC and Hitachi plants. Samsung and Fujitsu also increased the production capacity of existing plants. In 1998 Sanyo, Sharp, and Matsushita Electric started operation of new plants. Korean makers are also advancing rapidly; LG Semiconductors started operation of a new plant in November 1997. New plants of Samsung and Hyundai, expected in operation in 1998, can churn out 120,000 and 180,000 panels per month, respectively.

Also directly linked to the notebook PC market are batteries and hard disc drives. The worldwide production of lithium ion batteries has shown a sharp rise since their introduction in 1992. In 1996, yearly production increased to 116 million cells, 270 percent over that of the previous year.

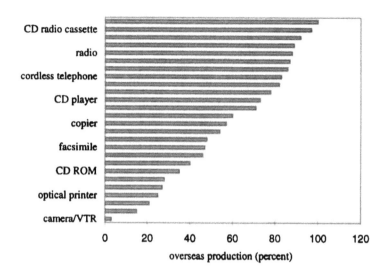

Figure 4.6 Histogram showing the percentage of overseas production by Japanese makers for various consumer electronics items as of 1995.

About fifty percent of lithium ion batteries go to notebook PCs, and another fifty-eight percent to portable telephones. Sanyo leads the worldwide market as of 1997, supplying batteries to Fujitsu, Nokia of Finland, and Siemens AG of Germany. SONY Energy Technology has been competing with Sanyo for market leadership. A technical factor of key importance is the thickness of the battery; the goal is to make notebook PCs ever thinner. In 1997 Mitsubishi Electric started shipment of "Pedion", a notebook PC with an overall thickness of 18mm that uses the lithium-polymer battery of UltraLife Batteries, Inc. of the U.S. The drawback of the lithium-polymer battery is its short life; it affords a mere one hour of operation. Battery suppliers are working hard to reduce the thickness of lithium-ion batteries while keeping the advantage of long life.

The hard disc drive market has been dominated by U.S. manufacturers like Seagate Technology, Western Digital, IBM, and Maxtor. Toshiba has a four percent share in the world market. Fujitsu and Samsung of Korea are driving their shares upward.

4.5.5. Telecommunications

Production of telecommunications equipment has exceeded that of computers in monetary terms since 1995, as shown in Figure 4.7 [*Nikkei*

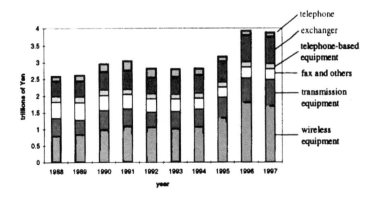

Figure 4.7 Histogram showing telecommunications equipment industry dominance

Electronics No. 675, 1996]. What promoted the growth of production in the telecommunications sector is the demand for wireless equipment and associated capital equipment (transmission equipment and exchangers). The year 1996 saw substantial investment by network providers in the infrastructure for mobile telephones — NTT Docomo appropriated 850 billion Yen, IDO 130 billion Yen, and Cellular Group 200 billion Yen. These figures are twice those of 1995. Industry watchers observe that, in the telecommunications sector, Japan is trailing the U.S. by about three years, with a projected annual growth rate of ten to fifteen percent toward the end of this century. Figure 4.7 shows a temporary halt of growth in 1997, due to concentrated infrastructure investment in 1996 and a lag in the rate of growth of the number of portable telephone users. Still, in the long run, one of the highest growth rates is expected in the telecommunications market.

The number of portable telephone users increased at a monthly rate of eight hundred thousand to a million in 1996. In order to keep up this pace, some visible incentive is needed, such as drastic reduction in telephone charges. Such incentive may be relatively easy to achieve, because the network providers run on their own capital infrastructure, free from reliance on the NTT telephone network. Another type of mobile telephone, personal handy-phone system (PHS), is designed for use in homes. PHS phones are popular among teenagers who enjoy late night chats with friends from their own rooms without interference from parents. The statistics of DDI Tokyo Pocket Telephone, a PHS provider, show that PHS usage was concentrated in a three-hour band from 9 pm to 12 am. PHS system providers and phone makers are working hard to cultivate another group of users; their current target is housewives. Currently, the telephone charge for PHS is inexpensive compared with mobile phone service, but the providers are paying about half of their sales revenue to the fundamental network provider, NTT.

In addition to the growth of mobile telephone users, the increase in Internet users and business intranets are contributing to expanded demands for exchangers. ISDNs for the Internet require enhancement of the exchanger capacity. Intranet-related demand for ATM exchangers increased by thirty percent in 1996 from that in 1995. NTT plans to restructure the communications infrastructure that will extend to 2010. One of the core projects is replacing copper wires in residential telephone lines with optical fibers. In December 1997, NTT announced that the plan to adopt digital-mode transmission was complete for all Japanese telephone lines.

4.5.6. Components and Packaging

Miniaturization of components and systems is where the Japanese have proved most adept. Their penchant for this technology is a great asset in the electronics age. Portable computers, cellular telephones, camcorders, and digital cameras — all are now targets in the relentless pursuit of compact packaging. Fierce competition among Japanese manufacturers to reduce the size of consumer electronics seems a replay of the calculator war of the 1970s. The weight of camcorders illustrates this intense competition. In response to a 1,800-gram model from Sharp that was released in 1992, SONY developed a 930-gram model in 1993 [Boulton and Pecht, 1994], then reduced the weight further to 660 grams in 1996 [*Nikkei Electronics No. 663*, 1996]. The thickness of a sub-notebook PC is another indicator of the tough competition. While most notebook PCs range in thickness from 43 to 50 mm, Toshiba's 1997 mini-notebook "Libretto" was only 34mm thick. Later in 1997, Mitsubishi Electric shocked the PC industry with its 18mm-thick ultra-thin "Pedion".

Miniaturization of components is one of the key ingredients in reducing system size. Resistors and ceramic capacitors with an exterior size of 1.0 mm x 0.5 mm (1005) and 1.6 mm x 0.8 mm (1608) are now commonly employed in portable equipment [*Nikkei Electronics No. 663*, 1996]. Further miniaturization to 0804 (0.8 mm x 0.4 mm) and 05025 (0.5 mm x 0.25 mm) is being attempted [Boulton and Pecht, 1994]. A dramatic reduction in the volume of a voltage-controlled crystal oscillator by TDK, ninety-six percent in seven years, is another example [Okamoto, 1994].

Another key ingredient for system-level miniaturization is the reduction of CPU-board and concomitant packaging technology that allows dense placement of chips and other components [*Nikkei Electronics*, 1997]. Flip-chip bonding of bare dies to build-up PCBs is being pursued by notebook PC manufacturers. The build-up PCB provides a 120 micron wiring pitch, well within the range of the LSI pad pitch. Build-up PCBs are becoming increasinmgly available. Also, the technology for bump connections between the die and the PCB has been developed, and the supply of underfill resins is increasing. The cost of build-up PCBs dictates that the overall size of such fine-pitch PCBs is limited to around 30mm x 30mm; hence, the build-up PCB serves as a wiring substrate for multi-chip modules (MCM). For example, NEC employed a build-up PCB for a double-sided MCM on which RAMs and graphic controller chips were mounted. An area on the motherboard occupied previously by separately packaged chips is reduced by three-quarters by the introduction of MCM.

Flip-chip bonding is becoming increasingly advantageous for portable PCs when the I/O pins per chip exceed three hundred. Gold bumps, coupled with solder electrodes or conductive adhesive electrodes, are becoming popular features in flip-chip bonding. Process equipment vendors and materials suppliers, such as Toray and Showa Denko, are working with PC manufacturers to advance flip-chip bonding. The next challenge for bonding and PCB technology will be a response to the reduction of the LSI pad pitch to 80 μm. Current technologies have proven commercially feasible for pad pitches above 100 μm.

For low pin-count LSIs, Chip-on-Board (COB) technology is becoming a popular bonding scheme, where bare dies are wirebonded to the substrate. An alternative to the use of bare dies is chip-scale packaging (CSP), in which packages have almost equal exterior sizes to those of bare dies. COB and CSP are being used in camcorders, cellular phones, cameras, and other consumer products. The production of PCBs with five to nine layers has the highest growth rate of all PCBs, with annual growth rate at sixteen percent [Nikkei Electronics No. 663, 1996], reflecting the demand for portable electronic equipment.

4.5.7. Materials and Process Equipment

Japan is importing poly-crystalline silicon from the U.S. (1,954 tons in 1996), Germany (800 tons), Ukraine (202 tons), Russia, and the U.K. Poly-crystalline silicon is further purified, then converted to high-purity, single-crystal ingots; wafers are sliced out of ingots. The process from crystal growth to wafer production is being performed by a number of Japanese manufacturers. The top seven are Shin-Etsu Semiconductor, Sumitomo Si-Techs, Komatsu Electronics, Mitsubishi Materials, Toshiba Ceramics, Tokuyama, and High-Purity Silicon. Shin-Etsu is a leader in the world market.

In spite of decreasing capital investment by chip makers due to the DRAM glut through 1996 and 1997, the volume of wafer supply showed no sign of adjustment [Nikkei Microdevices, 1996]. Suppliers increased production capacity to three million wafers per month at the end of 1996. In particular, epitaxial wafers are in large demand for 64M-bit DRAMs and 32M-bit flash memory chips.

A study is in progress to expand the wafer diameter to 300 mm by the year 2000 in mass production lines. The Semiconductor Leading Edge Technologies, Inc. (SELETE), co-founded by ten Japanese semiconductor makers, and the Japan 300 mm Semiconductor Technology Conference (J300), which provides a channel for information exchange and standardization studies, are coordinating the research and development covering the whole spectrum of agenda for transition to 300 mm wafers.

The first generation of subquarter-micron design rules (0.25 - 0.8 μm) will most likely be implemented on 300 mm wafers. Further ahead are emerging fabrication technologies for Gigabit DRAMs. For 1G- and 4G-bit DRAMs, the design rule needs to be reduced to 0.18 μm and 0.15 μm, respectively. Japanese chip makers are directing research and development efforts to deep sub-micron processing technology. The primary technical agenda items are enhanced self-aligned structures, rapid thermal processing, plasma doping, shallow junction formation, and high aspect-ratio contacts [Peters, 1997].

4.5.8. Electronics Manufacturing Services

Electronics manufacturing services have taken much of the advantage away from Japanese manufacturing firms through economies of scale in purchasing, logistics, and production. As shown in Figure 4.8, sales in the electronics manufacturing services (EMS) industry grew from $2.4 billion in 1984 to a projected $30 billion by the year 2000. In an IPC market study, communications accounted for 30 percent and the computer industry accounted for 43 percent of industry revenues as shown in Figure 4.9. Additional revenues were received from consumer/business retail (5 percent), automotive (3 percent), government/military (2 percent), industrial (13 percent), and instrumentation (4 percent). Companies over $175 million accounted for 88 percent of all revenues received from these markets. Since the computer industry was one of the first to embrace outsourcing and pay attention to margins and pricing, it is no surprise that it is the industry's largest customer. Further growth in outsourcing will lead to better representation of all industry segments.

Contract assemblers evolved into the electronics manufacturing services industry as they took responsibility for the final quality of the products being delivered to their customers. The EMS companies began as low-cost board stuffers. With the advent of surface mount technology, new investments in technology–both new equipment and educated technologists–original equipment manufacturers started to outsource rather than make the required investments. To overcome problems with sub-standard parts, EMS companies offered to source the PWBs and components and take responsibility for the quality of the parts provided. In addition to solving the quality issues, EMS companies added revenue by distributing and purchasing components and boards. In an IPC study, 86 percent of PWBs and 68 percent of components were purchased for the customer. To keep costs in control and processes efficient, EMS companies took over design or carried out comprehensive design review. Today, EMS companies offer full system building capabilities. Taking advantage of their lower costs and manufacturing expertise, EMS companies placed the completed assembly in

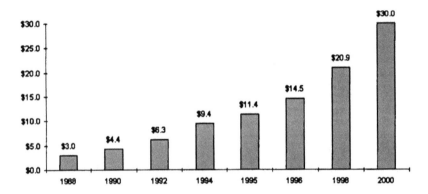

Figure 4.8 Revenue Growth for the EMS Industry.
Source: IPC

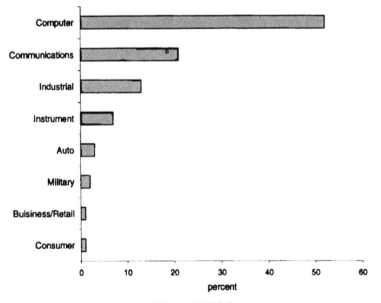

Figure 4.9 EMS Customers
Source: IPC

a product box. In 1995, 22 percent of the revenue came from actual assembly activity, 6 percent from design, testing, and repair and rework, 9 percent from other, and 63 percent from the sale of components and PWBs. Now they store the final product or ship directly to the OEM distribution center or customer. The IPC study indicated that 17 percent of revenues were reported to come from system building for larger EMS firms.

4.5.8.1. EMS Production Technology

Surface-mount technology is the core of the electronics manufacturing services industry. IPC's study participants reported 79 percent of components placed were surface-mount or fine-pitch components in 1994. And, although chip scale packaging is justifiably getting a great deal of attention, 74 percent of the components are standard surface-mount, 4.6 percent were fine-pitch, and the number of chip-mounted packages was under 0.5 percent. Out of the percentage of components placed, companies reported that 82 percent were surface-mount components in 1995, an increase from 77 percent in 1993. Of the surface-mount components, 79 percent were for surface-mount components greater than 0.25 mils, 16 percent were for fine-pitch (0.18 to 0.25 mils) and 6 percent were for ultra-fine-pitch (less than 0.18 mils). Less than 1 percent were for chip-mounted components. Companies reported that 12 percent of boards were assembled using hand assembly. Due to the large production volumes and greater equipment investment, companies with sales over $175 million reported only 6 percent in hand assembly. Companies under $5 million reported an average of 37 percent in hand assembly.

As the *IPC National Technology Roadmap for Electronic Interconnections* notes, "Although the high I/O components are a small part of the total by the numbers used, they are a big part of the factor that drives industry infrastructure in both bare board and assembly manufacturing." According to the Roadmap, high-pitch-count I/O components are growing at a very high average annual growth rate, between 30 percent and 70 percent. However, even with these growth rates, the Roadmap forecasts total usage in the year 2000 at only 6 percent of all components. With limited growth in through-hole components, standard and high-I/O surface-mount packages will dominate assembly technology.

Ball grid array packages are the "new workhorse of the high I/O industry," according to the Roadmap. Due to the array format, very high I/O counts are available without the drawback of bent, broken, or damaged leads. IPC forecasts increasing I/O counts on BGAs over the next five years, with counts as high as 1000 I/O. Additionally, the Roadmap forecasts replacement of quad flat-pack packages, especially those with over two hundred leads. Due to high assembly yields, solder self-centering characteristics, and large ball pitch, the IPC forecasts continued acceleration of BGA packages.

But EMS firms focus on regular, not cutting edge, technology. EMS firms are not research and development intensive. Their advantage is in high-to-mid volume, high-quality, efficient production. The investment needed to produce high-volume, cutting-edge products with lower yields does not justify the investment. However, some EMS providers have invested in prototype facilities to explore newer technologies [Sterling 1996].

4.5.8.2. EMS Competitors

Total EMS revenues grew from $11.4 billion in 1995 to $13.7 billion in 1996. With the assembly market estimated at $62 billion, the share for independent EMS companies was 19 percent. The top forty companies account for 74 percent of EMS industry revenue; the top twenty-five companies account for 63 percent of industry revenue. EMS companies are buying captive facilities to improve their own capacity and geographic coverage, both domestically and internationally. J.C. Bradford & Co. noted seventeen such acquisitions in 1995 alone, representing 1.6 million square feet. J.C. Bradford found average 7 percent return on sales in 1995 and $679 in sales per square foot. The IPC predicts 20-25 percent compound annual growth rates through the year 2000 for the EMS industry.

4.6. Packaging Portable Computers — An Example from Toshiba Libretto

The development of the Libretto series of portable computers by Toshiba engineers serves as an example illustrating the work of Japanese packaging engineers. In the spring of 1996 Toshiba began shipping the first model of the Libretto series, "Libretto 20," pronouncing it the smallest and lightest portable computer in the world among those with Windows 95. Its exterior size is 210 mm (width) x 115 mm (depth) x 34 mm (thickness), and it weighs 840 grams [Kamikawa 1997]. Its key features are an Intel 486DX4-75MHz microprocessor, a 270MB hard disc drive (HDD), a 6.1-inch thin film transistor (TFT) color display, a lithium ion battery good for three hours continuous operation, a keyboard with a 13 mm key pitch, a pointing device, a PC card slot, and an infrared communication port.

One target presented to the design team was to allow salary-men (white-color workers) to carry a portable computer in the inner pocket of the jacket. The first task was to survey the size of the pocket. The team sampled the pocket size from 1,043 jackets worn by salary-men. A critical measure was the sum of the height and depth of the pocket. Among the jackets surveyed, 200 had a height/depth sum of 151 mm, 180 had 149 mm, another 180 had

147 mm, 160 had 153 mm, 140 had 155 mm, and the rest ranged below 145 mm or above 157 mm. The team concluded that a computer with a height/depth sum of 149 mm could be accommodated by more than seventy-five percent of salary-men's pockets.

The next step was to build mock-ups that met these requirements. The crucial items were the size of the liquid crystal display (LCD) and the size of the keyboard, which defines the key pitch. There were two candidates for the LCD size (5.7 and 6.1 inches), and three candidates for the key pitch (11, 12, and 13 mm). The design team evaluated the mock-ups from every angle, and set the external size at 210 x 115 x 34 mm. The LCD size was determined to be 6.1 inches, and the key pitch was 13 mm. The overall weight target was 800 grams.

Three issues had to be determined at the beginning of the system packaging design. The first was the locations of the HDD and of the PC card slot. Based on the weight balance, the HDD was located in the left zone of the system box, and the PC card slot in the right. The second issue was determining locations of the expansion connectors. Considerations of available space, assembly cost, and structural strength favored putting the expansion connectors on the bottom of the system box. The third issue was the direction of battery insertion and removal. Closely related to this issue was the geometry of the battery and the size of the connector. The rectangular geometry of the battery dictated that, if the battery was slid into the system box from right or left, the connector had to be made small, and the small connector raised concerns about possible damage in the event of a short circuit. Safety precautions prevailed, and the battery was attached to the system box from the front with widely-spaced connector pins.

Due to the shrunken space in the system box, five components had to be reduced in size: the ASIC chip set, the HDD, the keyboard, the PCB, and the LCD. In conventional design the circuits for graphic and communications control are implemented in the chip set. In Toshiba's earlier designs the chip set consisted of five ASIC chips. For Libretto, those control circuits are implemented on a single ASIC chip, reducing the package volume by one-eighteenth of the previous volume and the package area by two-thirds.

The 2.5-inch HDD originally had a thickness of 12.7 mm. For Libretto the HDD thickness was reduced to 8.45 mm. The newly developed thin HDD proved its commercial value on its own. Competitors such as SONY came to purchase Toshiba's thin HDDs for their notebook PCs.

The keyboard is given a full set of eighty-eight keys, compatible with OADG (Office Automation Design Guide) 106. The key pitch is 13 mm, the stroke is 1.5 mm, and the total thickness is 6 mm. A wider key pitch and fewer keys were considered, but compatibility with the OADG 106 was given top priority.

The newly developed PCB has six wiring layers and is 0.8 mm thick. The manufacturing group worked out a scheme to mount a Pentium TCP with 384 pins and a 0.25-mm pitch on the PCB. Other components on the PCB had to be redesigned in order to accommodate them. The most significant impact on the PCB area economy was brought by direct mounting the HDD on the PCB, which reduced the PCB area for other components further, to one-third of the area in previous models.

In designing the LCD, a compromise was reached between the panel size and the thickness. The adoption of 6.1-in. display necessitated folding the TAB portion to the back of the panel, making the entire display thickness 7 mm.

Windows 95 capability strained the power resource and the thermal design. The CPU chip consumes 2.5 W, while the standard lithium ion battery has a capacity of 1200 mAh. A rule-of-thumb calculation, assuming 3.3 V for the operating voltage, indicates that the battery allows only 1.6 hours of continuous operation. However, an elaborate power-saving design elongates the period of continuous operation to two to three hours. This design consists of seven stages of power supply, from full power through high power, medium power, low power, suspend, hibernate, and shut-down. The power supply stage is determined by monitoring the job load and the residual capacity of the battery.

Tight constraint on the space excludes any possibility of providing internal routes to vent air from the system box, let alone installation of a cooling fan. Heat from the CPU chip and other devices is conducted to the box enclosure, then dissipated to the atmosphere by natural convection and radiation heat transfer. A challenge arose in the design of heat spreaders to be mounted on the CPU chip (encapsulated in a tape chip carrier (TCP)). In earlier designs, the TCP was sandwiched between two metal heat spreaders. To reduce the height of the assembly, the metal spreaders were removed; instead the PCB was given the function of a heat spreader. To enhance heat spreading, the ground and power layers are made thicker than those in conventional PCBs (105 μm) [Happoya, Takahashi, Igarashi, and Nishimura, 1997].

It took almost a year to develop the first of the Libretto series (Libretto 20) [Record of Round Table Discussion, 1997]. As always in the case of newly released models, the majority of first purchasers were technology enthusiasts. Their response was important for the next round of development. It happened that many customers were not satisfied with the small user space, which resulted from the installation of Windows 95. Some of those users modified the product on their own, cutting out the bottom of the system box and installing larger capacity HDDs. About seven months after the introduction of Libretto 20, Toshiba responded to the complaint by releasing Libretto 30, which had an enhanced HDD capacity of 500 M Bytes.

The clock frequency was raised to 100 MHz. Then, two months after the release of Libretto 30, the next generation model, Libretto 50, was introduced. The major modifications in Libretto 50 are the use of Intel Pentium for CPU (75 MHz), enhancement of the HDD capacity to 810 M Bytes, and the use of magnesium alloy for the system enclosure.

A magnesium alloy (magnesium with nine percent aluminum) was used, skimming the enclosure material by at least 0.5 mm and creating an additional 1 mm in the system box. The Libretto 50 required this additional space to accommodate the components with enhanced functions. A common enclosure material for portable computers has been plastic, which affords a minimum thickness of 1.2 to 1.4 mm. As an alternative, the magnesium alloy was evaluated. The usual method of forming metallic enclosures has been die-casting; however, with die-casting, the thickness could not be reduced below 1 mm. Toshiba engineers turned to injection molding, which was being employed to form the chassis of video and other equipment. In injection molding, a paste of metal powders and binder is fed to the molding machine. The expertise to reduce the enclosure thickness to less than 1 mm by injection molding is limited to only a few manufacturers, including Toshiba.

After Libretto 50 Toshiba developed Libretto 60, then Libretto 70, in about a three-month period. The clock frequency was raised to 100 MHz in Libretto 60, and to 120 MHz in Libretto 70. The metallic enclosure apparently also assisted the dissipation of the greater heat associated with increasing the clock frequency.

The development of a series of new models in so short a period was possible only through close cooperation between engineers from a wide spectrum of disciplines. Mobilization of a wide range of human and technical resources is relatively easy in the hierarchical organization of a big company group like Toshiba; Japanese group conglomerates will be discussed in the next section.

Unlike their predecessors in the catch-up phase of the 1940s and 1950s, engineers of the present generation are fully informed about current Western technology and have almost every technological resource at their disposal. They no longer have to compete for Western knowledge to stay ahead of their colleagues and competitors. Yet, they work hard. The motivations that drive them to long work hours are several. First, the corporate atmosphere instills a conformist mentality. Everyone works long hours. Secondly, the joy of materializing one's ideas into real products is very rewarding. The tools of design and prototyping have become so sophisticated that those who master the art of computer-aided design or operating manufacturing equipment can play on their own ideas. In the old days young start-ups did chores for their seniors in exchange for on-the-job training. This master-disciple relationship was well in tune with Asian culture. It still exists, but its intensity is diluted because of the greater independence junior engineers enjoy due to the sophistication of engineering tools. For instance, today a senior engineer has to rely on the computer simulation run by a junior colleague, whose work remains opaque to him. The danger of overlooking some important points is involved; however, hidden errors, if any, may surface at a later stage of prototype testing. More important is the sense of independence imparted to junior engineers.

The development of consumer products such as Libretto obviously benefits from the initiatives taken by young engineers. Unlike the development of mainframe computers, the development of portable equipment does not require large-scale engineering and bureaucratic design screening. As technology is heading toward the era of mobile computing, there will be growing demands on portable electronic equipment and more room for young Japanese engineers to play active roles. In the years of the bubble economy of the late 1980s, concerns arose among Japanese industry leaders and educators about the dwindling popularity of the engineering profession. That concern proved almost needless as the nation now finds young generations enjoying the work through desktop computers and numerically controlled manufacturing machines.

4.7. Metrics for Future Electronics Products

Unlike most technologies, electronics products encompass a wide range of technologies. The primary trade-offs in supporting electronics technologies include (1) size and weight dimensions, typically described in millimeters and kilograms; (2) functionality and performance, most often described in terms of features and speed of performance, and energy consumption; (3) shape; and (4) the final cost of the product or its key components.

4.7.1. What Are the Overall Size and Weight Requirements of the End Product?

SONY has traditionally had the objective of introducing portable consumer products like the Walkman, Handycam, and Discman. Each generation of new products has been smaller and lighter in weight. For example, SONY's first model of Handycam, released in 1985, weighed 1,970 grams. By the mid 1990s, the weight was under 800 grams, a 60 percent reduction [Kaneda 1994]. Consumer products which use optical pickups, such as the compact disk (CD) players and mini disk (MD) player/recorders, get their digital signal from its laser diodes which are only 0.2 x 0.25 x 0.12mm in size [Kaneda 1994]. The laser diode requires ultra-precision mounting equipment to place the laser chip on a silicon wafer. Assembly requirements for such miniaturized components are the most advanced available. Today, cellular phones are driving Japan's miniaturization efforts. Cell phones, averaging over 1100 cubic centimeters in 1985, shrank to 106 cubic centimeters in SONY's 1995 models, a 90 percent reduction in size in less than a decade. Component suppliers like TDK have reduced the volume of its (VCO) for use in cellular phones by 96 percent in seven years, down from 26 x 17 x 10mm to 10 x 7 x 2.5mm [Okamoto 1994]. Reductions in energy consumption through miniaturization allow smaller batteries to be used.

The continued reduction of product size and weight has pushed suppliers to develop and introduce smaller components and related high-precision assembly equipment. Once the natural limits of miniaturization are reached (e.g, the Walkman or Discman approximates the size of a cassette tape or compact disks) smaller component formats are introduced (e.g., mini tape cassettes or mini compact disks) to allow for further miniaturization and increased portability. For example, Sharp introduced a four-pound 8 mm ViewCam camcorder in 1992. SONY's response was to introduce a two-pound model in 1993. Except for cellular telephones, few U.S. manufacturers have size or weight limitations that are as demanding as portable consumer products. In fact, while Japanese consumers demand much smaller products than most Westerners, they are beginning to complain that cellular phones are now too small, easily lost in a woman's purse or too short to reach from the ear to the mouth for clear communication.

4.7.2. What Functionality and Performance is Required of the End Product?

While advanced products like supercomputers and satellites have traditionally driven electronic performance and function, they are typically too expensive for most consumer product applications. However, consumer products are requiring more advanced technologies and performance–global positioning systems, mobile satellite communications equipment, high-definition TV, or fuzzy logic controlled microwave ovens. The move into multimedia products will demand ever-increasing functionality, integrating multiple technologies. For example, Sharp's ViewCam introduced a teleport for sending still pictures over analog telephone lines to other ViewCams. Sharp's Zaurus uses infrared to send facsimiles, communicate with other computers, or print. Next-generation products can be expected to integrate more combinations of technologies into single products, like Sharp's combined TV/VCR sets. The new technology produces unheralded quality in graphics and sound. Once such integration occurs, the process of miniaturization and weight reduction will continue. With increased functionality also comes the demand for increased performance. For example, we see the growth in software capability driving the demand for higher performance computers with faster processors. Sega introduced its Dreamcast game machine in 1998 using a next-generation 128-bit microprocessor produced by Hitachi.

4.7.3. What are the Shape Requirements of the Consumer Products?

Consumer products also require advanced styling, such as we see in today's cameras. The Japanese dominance of the single reflex and video camera markets has led to the development of advanced flexible circuit boards that mold to the shape of the camera body. Canon, Minolta, and Nikon cameras all use complex flexible circuit boards. Products like Sharp's new ViewCam also utilize flexible connectors to tie together multiple boards for use in miniaturized video systems. Sharp is currently developing a multi-layer flexible PWB for its advanced product development requirements.

4.7.4. What are the Cost Requirements of the End Product?

The slim margins from final assembly have hollowed out many industries as companies move to low-labor-cost countries. Japan has lost its leadership in low-cost consumer markets as manufacturing has increasingly been outsourced. Typically, key components used in electronics products amount to over 50 percent of the total cost of the product. To cut component costs, the number of parts must be reduced. SONY cut the number of parts in its optical pickup from eight to two in 1994, drastically cutting both its weight and cost. Innovative designs and high levels of electronic integration can cut costs from 30 to 70 percent [Kaneda 1994]. Both design and integration methodologies reduce the number of parts needed for assembly.

4.8. Scenarios for the Future

The Japanese electronics industry seems self-contained in the sense that all basic materials, components, and manufacturing technologies are available in the domestic market, except for Intel's microprocessor chips and Microsoft's Windows. The fact that these two vital elements of the electronics industry, the microprocessor and the operating software, are in the hands of American companies has irritated bureaucrats, industrial leaders, and intellectuals. The leading newspapers and magazines occasionally carry articles that lament the incompetence of Japanese scientists and engineers in inventing such key technologies. In a recent interview in the *Mainichi Shinbun* [*Mainichi Shinbun*, 1998] a well-known theater scriptwriter went beyond his professional arena, saying that Japan should spend more on fundamental scientific research in order to produce such breakthroughs as

Intel's microprocessor. Although the microprocessor and Windows did not come directly from fundamental research in science, such rallying cries moved the government to allocate greater budget resources to scientific research. While this will be discussed further in the next chapter, this account is meant to illustrate the common mentality held by the Japanese — that is, every vital thing has to be manufactured in Japan. This mentality is further narrowed down to the corporate level — that is, every vital thing has to be manufactured in one's corporate group.

Toshiba Corporation heads the Toshiba group, which covers almost all engineering disciplines. The group is a typical Japanese conglomerate. Many big Japanese corporate brands, such as Hitachi, Matsushita, Mitsubishi, Sanyo, SONY, Sharp, and Toshiba, represent a group. Strong human ties are cultivated within the group by personnel transfers within the group. Almost all top management positions of the subsidiary companies are occupied by former higher-ups of the lead company. Also, middle managers are transferred from upper-level companies to lower-level companies and compensated with promotion to the next higher rank; for instance, section manager in an upper-level company to division manager in a lower-level company. This arrangement has been in place for almost fifty years, since the end of World War II. The procurement of materials, components, and services, and the transfer of technology are all lubricated by these personal ties. When all basic materials and technologies are available within a single corporate group, the development of new products is not only easy but kept from the sight of competitors until the moment of product release. This gives a sense of security. Any elaborate legal arrangement among participants in product development to protect proprietary information is unnecessary.

Consequently, at the national level it has been assumed that the dependence on foreign companies in the area of vital technology is dangerous. What will happen if the foreign suppliers decide to withhold the supply? The government bureaucrats view this possibility as a hidden threat to the national security. This mentality is commonly held by bureaucrats of many other nations, but in Japan, the hierarchical organization pervading the country made it easy to implement the bureaucrats' plan. Since World War II, MITI has guided the industry to cover all technological areas that MITI deemed vital. One area whose strategic values are easy to comprehend is materials processing. Starting with steel making, the production of other metals, resins, ceramics, silicon crystals, and so on have been entrusted to the companies assigned by MITI. These assignments facilitated the flow of capital to import the technologies, then refine them. The result is the establishment of specialized companies on a scale unseen elsewhere in the industrialized countries [Clark, 1979]. Although their names are hidden from general consumers, the materials processing companies are now among the mightiest companies in the world, in the sense that they control the flow of materials. In the electronics industry particularly, Japanese companies are supplying basic materials worldwide: silicon crystals, leadframe metals, heatsink metals, packaging resins, cables, optical fibers, and others. Their shares in the world market are staggering, with some reaching fifty to ninety percent. In July 1993, the fire at Sumitotmo Chemical's resin plant panicked semiconductor manufacturers worldwide. This episode made the public aware that this plant on Shikoku Island was vital to the semiconductor industry.

As of 1998, in the face of the Asian economic crisis, the whole organization of Japan's economy is under question. The domestic economy has stalled. With organizational expansion being halted in all industrial, business, and government sectors, the descending flow of personnel from upper-level organizations to lower-level organizations has become sluggish. Instead of trying to overhaul the system, salaried workers are concerned with a more immediate issue — how to secure a job after retiring from an upper-level company. The retirement age for middle managers at top-rank companies is fifty-five; obviously, many still have to work to continue paying house loans and sending their children to school. Not only those leaving jobs at upper-level organizations suffer. Worse situations await those who started their careers at smaller companies. Most small companies depend on the domestic market; hence, they are vulnerable to an economic downturn. Job security concerns make people defensive, reducing their spending further.

Competition at individual and corporate levels must intensify to secure the future. Intense competition often favors the expansionists in the company rather than the moderates. With the domestic market frozen, expansion is directed to foreign markets, in particular, the U.S. market, exacerbating the trade imbalance between Japan and the U.S. There is no end in sight to the trade controversy that has been smoldering for decades.

A pessimistic scenario for the future of Japan is worth consideration. In the worst case, Japan could become more closed than ever to real exchanges with other parts of the world. On paper, progress would be made in dismantling various measures to protect domestic industry from foreign competition and in lowering tariffs on imported goods. At the same time, real barriers will grow sky-high, as internal arrangements within the corporate group, or within Japan at large, supersede the free market. Intra-group trade will be preferred to procuring cheaper materials and components from external sources. Such preference is partly motivated by job security. Anyone planning to step down to a subsidiary company in a few years will not want to intimidate the current holder of the job by canceling purchase orders to that company. The hierarchical organization will become more vertically oriented, rigid, and impervious to external influences.

The self-sufficiency mechanism will survive as younger generations find the environment created by the mechanism cozy and convenient. Young engineers will become more insensitive to events in the external world, although they work diligently perfecting miniaturization and manufacturing. The quality of made-in-Japan goods will not degrade as future engineers continue to attend to details to remove imperfections from products. Globalization will continue and expose more engineers to work experiences in overseas manufacturing plants and service facilities. But such experiences do not necessarily produce internationally-minded engineers. Often, overseas plants are mini-Japan, stuffed with Japanese goods, foods, newspapers, and magazines.

Those entrepreneurs who founded the consumer electronics industry in Japan have already left or are now leaving the scene. The industry has now matured enough that it no longer needs strong personalities for further development. It still has many innovative engineers, but they are incorporated into the hierarchical organizations.

While people will become more inwardly oriented, manufactured goods and basic materials will continue to attract foreign buyers because of their high quality. Japanese brands will continue to flood the world, but Japanese industry will become more faceless due to the lack of visible personalities.

If the future follows this scenario, the Japanese will suffer most. The aging of the population forces the nation to reassess the assumption that the present way of earning through manufacturing will work well indefinitely. By the year 2020, more than one in every four Japanese will be sixty-five years or older. Even now, the pensions in Japan are the lowest among industrialized nations, and some pension plans of small companies are bankrupt. The government's policy has been to balance the national budget. It raised taxes for this purpose in 1997, the worst time ever, chilling further an economy already at low ebb after the bursting of the bubble.

Jobs must be created for the aging population to alleviate the burden of paying pensions. As the economy is activated by the expansion of work opportunities, the resources for pensions will also grow. With less concern about their future, people will relax, lose the everything-Japanese mentality, seek lifestyle diversity, and create a large market for imported goods. The number of early retirees will grow dramatically as the baby boomer generation reaches its mid-fifties in a few years. Assuming that most manufacturing companies are more or less related to the electronics industry, it is estimated that each year in the early 2000s, around a hundred thousand professional engineers will have to leave the companies where they started their careers.

Where will job markets be created for these engineers, who have to leave their companies simply because of their age and cannot land new jobs as their predecessors did? One popular job opportunity is teaching in private universities. Today, it is common for one such opening to receive three to four hundred applications from those about to reach retirement age.

The knowledge and expertise of experienced engineers can be traded as a commercial commodity. Those "retired" engineers should be invited to remain as participants in the technological progress. Instead of letting their expertise go to seed, they would have to continuously update it through active participation in design and research work as part-time consultants, external design subcontractors, and so on. The problem is that this would be a new concept in Japan.

A crucial role in the creation of jobs for senior engineers will be played by educating the public on the value of abstract concepts and ideas. As described before, the Japanese historically have a high regard for novel ideas from the West, but domestically produced ideas were instantly diffused without due recognition going to their inventors. Today, patents are filed in record numbers, but few of them are outstanding. There are a number of similar patents based on apparently copied ideas with minor modifications. Kinship fostered in the hierarchical organization lets people assume that any ideas and knowledge belong to the group, with no need to reward those who toiled to produce them. When the university professor serves as a consultant to a company, the reward is usually a box of cakes or a pack of beer tickets or a dinner in the nearby restaurant. Japanese engineers visit American universities on a similar assumption that they may get new technical information in exchange for little souvenirs. American professors have come to expect that substantial financial support may ensue, but often find nothing on the end of the string. This lack of respect for abstract information is a conspicuous Japanese characteristic.

A strategy is clearly necessary to correct this cultural characteristic, create jobs for experienced engineers, and gain other benefits.

The Japanese government and the Diet should enact a law designed to encourage the founding and operation of small consulting firms in Japan. A tax exemption to manufacturing companies for the expenditures on subcontracting to consulting firms run by retired engineers may serve as an incentive. Also, a reduced charge for Internet and other networking services may encourage this kind of work. Coupled with this incentive should be the development of an information network infrastructure, in which Japan lags well behind other advanced countries. Measures to protect proprietary information must be tightened to educate the people about the commercial value of knowledge. After all, in the information age, knowledge is the most precious commodity.

The existing hierarchical organization and people's tendency toward conformity should be exploited for good. Once government-led measures take root, they will permeate throughout the nation. As people become used to buying knowledge, more incentives can be given to creative and innovative talents to formulate tradable knowledge that can find its way to export, helping to correct the enormous surplus of knowledge imported to

Japan from abroad. A diversification of the Japanese mindset will ensue as professional expertise receives higher regard. People will come to remember each other not by the name of their employers, but by their professions.

Professional engineers in the electronics industry may not constitute a major proportion of the work force; however, a crack in a small corner of a society may propagate throughout the whole society. Eventually, the mindset fostered by hierarchical organization will relax and Japan will gain full membership in the information community. This is a good scenario for the world as well. Existing electronics companies should help implement this scenario, because the sales of equipment and the development of the information highway infrastructure will also be good for their businesses.

A society gives birth to new technology out of its needs and on the basis of its infrastructure. The advent of new technology in turn changes the society. In order to precipitate any significant change in the society, we must rely on the dynamics between the technology and the society. In the case of Japan, government action must play on these dynamics to increase the nation's economic health.

Chapter 5

Research, Development, and Education

Criticism has been brought by foreign governments and other sources, claiming that Japanese industrial success has been built on patents and licenses bought from other advanced nations. The so-called "free-ride theory" has obsessed industry leaders, government bureaucrats, educators, and intellectuals in Japan. Calls for more Nobel prizes, eye-popping scientific breakthroughs, and technical inventions comparable to the invention of the transistor, are moving the government and corporate management to invest growing funds in basic research. The university system is also undergoing restructuring, with the former imperial universities gaining greater roles in fundamental research. In this chapter these current trends are reviewed, and their implications for the Japanese electronics industry are examined.

5.1. Macro-scale Perspectives

Figure 5.1 [*Science and Technology Indicators*, 1997] shows major suppliers of technology to the USA. Overall, in recent years Japan has fared well compared with European countries. Two factors are responsible for Japan's success — the commercial success of the Japanese industry and the content of the technology trade for Japan. As of 1995 about fifty percent of Japan's income from the technology trade came from Asia, thirty percent from North America, seventeen percent from Europe, and the rest from South America, Oceania, and Africa. Japanese payments in the technology trade are more biased toward North America and Europe — about seventy-one percent to the U.S. and twenty-eight percent to European countries. Major European recipients of payments from Japan are the U.K. (three percent), the Netherlands (six percent), Switzerland (five percent), Germany (five percent), and France (five percent). The statistics reflect a pattern: technologies originally come from the U.S. and western European countries, are refined in Japan, then transferred to Asian countries and also back to the U.S. and Europe, accompanying overseas manufacturing plants.

The development of new technology from scratch is a costly and time-consuming venture. It is imperative for Japan to shoulder part of the burden of originating new technology as it piles up a surplus of trade in

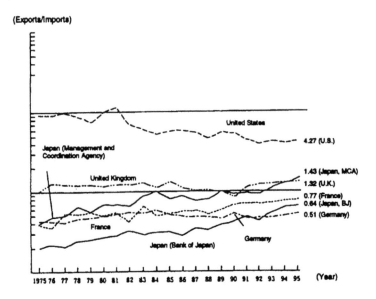

Figure 5.1 Graph shows major suppliers of technology to the USA [*Science and Technology Indicators*, 1997]

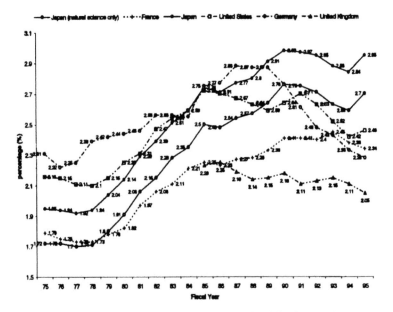

Figure 5.2 Graph showing the expenditure on research and development in major countries in terms of the percentage of GDP [*Science and Technology Indicators*, 1997].

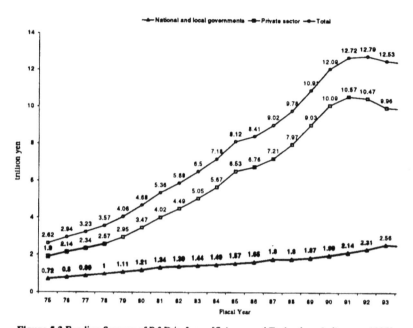

Figure 5.3 Funding Sources of R&D in Japan [*Science and Technology Indicators*, 1997]

manufactured goods. Several indicators show the nation's potential for leading science and technology: the expenditure on research and development, the human resources in science and technology fields, the statistics in knowledge transfer in the form of patents and publications.

Figure 5.2 shows the expenditure on research and development in major countries in terms of the percentage of GDP [*Science and Technology Indicators*, 1997]. There are two curves for Japan, one showing the expenditure on all science and engineering R&D, the other giving expenditures on natural sciences only. In the 1990s Japan has come to exceed all other nations in both indexes.

Figure 5.3 [*Science and Technology Indicators*, 1997] gives statistics on funding by the private sector and by the Japanese government. As of 1995, about seventy-eight percent of the financing came from the private sector, and twenty-two percent from the government. It has been repeatedly pointed out [Mukaibo, 1991, Itoh, 1995] that funding by the private sector has served the process of digesting and refining the technology from the West, and that comparable government funding has been meager.

It is widely believed that government spending on research and development is the key to encouraging the high-risk ventures in scientific investigation that lead to major breakthroughs. Figure 5.4 also shows the ratio of government spending on R&D to the total R&D expenditure in the

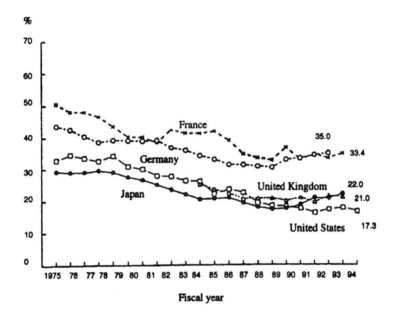

Figure 5.4 Graph showing ratio of government spending on R&D to the total R&D expenditure in the major developed countries (military-related R&D is excluded) [*Science and Technology Indicators*, 1997].

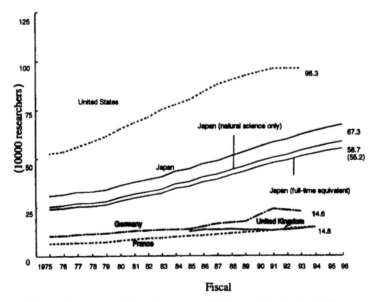

Figure 5.5 Graph showing population of researchers [*Science and Technology Indicators*, 1997]

10000 applications

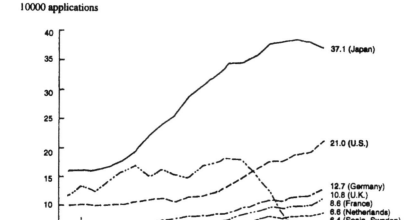

Figure 5.6 Graph showing number of patent applications [*Science and Technology Indicators*, 1997]

major developed countries (military-related R&D is excluded). Japan had been at the bottom of this index until 1991, but has come to exceed the U.S. and the U.K., thanks to the appeal to the government from academic and industrial leaders. Human resources for the Japanese science and technology also show impressive numbers. Figure 5.5 [*Science and Technology Indicators*, 1997] shows that the population of researchers in Japan is higher than that of any one European country. The U.S. has roughly twice the number of researchers as Japan, so that on a per-capita basis Japan is on almost equal footing with the U.S.

Japan is well ahead of any other foreign nation in the number of US patent filings, as shown in Figure 5.6 [*Indicators of Science and Technology*, 1997]. Figure 5.6 [*Science and Technology Indicators*, 1997] shows the growth in Japanese corporate patents registered in the U.S.A. Here again, Japan surpassed major European companies in the 1990s.

All these macroscopic indicators imply that Japan is going to dominate the science and technology of the next century, although it may take some years for the science policy initiated in the early 1990s to produce real effects. Before jumping to such a conclusion, however, we shall examine the myths believed by policy makers, industry leaders, academia, and the general public.

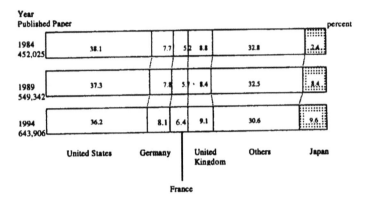

Figure 5.7 Graph showing the percentage of scientific publications per country in total world publications [*Science and Technology Indicators*, 1997]

5.2. Dynamics of Technology Development

In considering the roles of the Japanese electronics industry in the worldwide technological progress, we need to shed light on two technological areas. One is the technology that is needed to produce basic materials and components, such as integrated circuit chips. The other is the technology for innovation in consumer products, which depends on the synthesis of design, manufacturing, and intuition about consumer demand.

In basic technology, the Japanese electronics industry has excelled in incrementally extending established paths of technology. It has great inertia. For instance, Moore's law for integrated circuits has been faithfully followed by Japanese chip makers. This progress in circuit density looks impressive, but the basic content of the technology has not changed, no matter how high and how fast the density has been increased. Moore's law is based on improvements in the quality of silicon crystal and processes such as diffusion, etching, ion implantation, the precision of stepper movement, the level of clean room environment, and so forth. In these areas Japanese industry has achieved rapid progress.

When new technology emerged, however, the industry did not move as quickly. Japanese chip makers were slow in adopting flip-chip bonding in commercial products, although they established technical expertise in that area soon after IBM used the technology for mainframes in the late 1970s. A report on Japanese packaging technology [JTEC Panel Report, 1995] contains comments by Japanese engineers showing their reluctance to jump on new schemes, such as ball grid array (BGA) bonding. They preferred refining existing technology, maintaining that the extension of the technology is more economical than the switch to new technology. The placement pitch of I/O pins on the leaded chip carrier (existing technology) has been reduced to small values never realized outside Japan. However, once a switch of technology is effected by commercial incentives, the same great inertia pushes that technology rapidly. Flip-chip bonding and BGA technology are now being widely applied to chip and module bonding in portable computers.

In contrast to the attitude toward basic technology, consumer product development is characterized by remarkable flexibility and agility, a manifestation of characteristic Japanese entrepreneurship in consumer electronics development. Present-day entrepreneurs in consumer electronics are not visible personalities, but groups of engineers working in large companies. The development of Toshiba's Libretto is the product of one such entrepreneurial group. Streams of new consumer products — cameras, videos, computer games, portable phones, pagers, all with a variety of new functions — are being generated by invisible entrepreneurs. Consumer response is fed back to the development group through sales agents, traditional media like newspapers and journals, and web page questionnaires, reporting not only sales volume but technical details regarding performance and user friendliness. The late Akio Morita, co-founder of SONY, had been most vocal in pointing out to overseas observers this sensitivity to consumer demands by Japanese engineers.

The Japanese have unquestionably made contributions to commercial applications of electronics technology. This is where they excel, for both social and cultural reasons. The hierarchical organization encourages them to be innovative in the development of end products, where the reward of work is most visible. Constrained by tight space and by close community, the Japanese have come to please each other by exchanging well-crafted small items. These social and cultural factors should not be overlooked when developing fundamental science and technology.

The birth and subsequent development of new technology is a complex process involving a number of social, cultural, and technical factors, and the course of technological development is often determined by fortuitous turns of events. The government's policy of building the infrastructure to produce more technological breakthroughs, measured by the macro-indexes described in the previous section, is one essential incentive, but not all that is needed.

Historically, important social needs prompted the inventions that changed the world, and often, such needs involved large numbers of people. The rapid influx of immigrants to the U.S. in the late nineteenth century produced the prospect that the national census conducted in 1890 could not be processed in time for the next census in 1900. The U.S. Census Bureau needed a mechanical number-processing system, a need that prompted the invention of the Hollerith machine. In the 1930s the growth of telephones in the U.S. required large numbers of telephone operators. AT&T, then monopolizing the telephone industry in the U.S., feared that the required number of operators might reach half the U.S. population in years to come. The transistor was invented against this backdrop.

For an invention to grow to become industrial technology, it must have a financial boost. Electronics technology owed its initial growth to huge military spending by the U.S. at the height of the cold war. The U.S. military purchased almost fifty percent of the transistors produced throughout the 1950s. While selling products to the military, technology pioneers developed the expertise for mass production. Manufacturing capacity soon exceeded military purchase volume. The companies then turned their attention to possible commercial applications. In the late 1960s Intel's technical capability had matured enough that Intel could respond to an inquiry from a Japanese calculator manufacturer.

What big social needs do we have now? One of the popular subjects in Japanese academia, and to some extent among government bureaucrats, is the preservation of the earth's environment. Closely related to this subject is the conservation of energy. Japan is making substantial contributions to a scientific understanding of the deterioration process of the environment by providing the world community with data obtained both on the ground and from satellites. For industry, however, all that it needs to help preserve the earth's environment is incremental progress in making existing technologies less energy-demanding and in reducing the volume of material in products. Japanese industry has already made strides in conserving of energy in industrial and utility plants, and automobiles.

Other subjects take priority in the minds of the populace. According to a survey made by the Science Policy Laboratory of the Agency of Science and Technology, seventy-three percent of the respondents gave top priority to research in medical technology [Fukuda, 1997]. Also ranked near the top are disaster prevention technology and the preservation of the environment.

How can these needs be translated to scientific research programs with specifically defined targets? History tells us that any scientific research is bound to fail, or at least produce little impact on society, if it has no specifically defined target. Due to difficulty in defining the technical details of these research targets, government agencies resort to blending those general needs with the extension of existing technologies, with expectations for already known and highly publicized scientific findings, and with wild imagination. Table 5-1 shows the technical areas to be targeted by Japan's R&D on recommendation of the Economic Planning Agency [Lorriman and Kenjo, 1996]

The end of the cold war meant the end of big funding by the U.S. military for boosting fledgling technologies. Today's initial boosters come from the commercial sector, in the form of either in-house or external funding known as venture capital. Unlike military sponsors, commercial

Table 5-1 Technical areas to be targeted by Japan's R&D on recommendation of the Economic Planning Agency [Lorriman and Kenjo, 1996]

Expected to be number one in 2010	Expected to be among top players in 2010
Terabit memories	Super-intelligent chips
Superconductor devices	Self-breeding chips
Terabit opto-files	Opto-computing device equipment
Terabit opto-communication devices	Biocomputers
Biosensors	Super-parallel computers
Neuro-computers	Automatic translation systems
Super-conductor materials	Opto-ICs
New glass	Molecular devices
Hydrogen-absorbing alloys	Dementia drugs
Magnetic materials	Immunity/allergy drugs
Fuel cells	Bioenergy
Solar generators	Artificial organs
High-efficiency heat pumps	Video-conference systems
Intelligent robots	Videophones
AI-CNC	Wideband ISDN telecommunications
Hybrid machining centers	Opto-subscription systems
HDTV	Opto-LAN
Super-conducting linear vehicles	Next-generation automobiles
Next-generation superconductor linear vehicles	Innovative automobile manufacturing technology
Bimodal systems	Ultra-high multi-story buildings
Techno-superliners	Marine leisure lands
Intelligent ships	CO2 consolidation using catalysis
Gravitation-free underground laboratories	CO2 consolidation using plants
Linear motor catapults	Underground water reservoirs
Underground delivery networks	
Deep underground railways/roads	
Earth heat reservoirs	
Marine stock farms	
Fluorocarbon retrieving technology	

funding agencies demand quick returns on their investment. Hence, the paths of technology development now and in the future will not be the same as we have seen in the development of semiconductor technology. In Japan, the Ministry of International Trade and Industry is calling for the creation of a venture capital market. However, MITI's plan has not yet materialized, because the initiative has been left to the private financial sector, which has been in bad shape since the bust of the 1980s bubble. Because of the lack of specifically defined research targets and the absence of initial boosting mechanisms, it is highly likely that the present growth in funding for fundamental scientific research will produce nothing substantial to mitigate the free-ride theory. Still, some companies have established central research or fundamental research laboratories to emulate the Bell Laboratories of the 1940s.

However, this may not be the best — or only — route to take. Both the free-ride theory and the Japanese government's and industry's desire to generate new technology from Japan are based on a notion bound by past historical experience. It is widely believed that the transistor was invented by solid-state physicists, thus the next generation of technology will also originate from fundamental scientific studies. Such a possibility exists, but research in traditional scientific disciplines may not be the only source of new technology. True, the inventions of the transistor and the integrated circuit were engendered by an intellectual and technical infrastructure fostered by the advancement of fundamental science. However, the actual trigger which directly led to the invention was a synthesis of knowledge precipitated by a well-defined mission for research, not the advent of a novel science. The process of knowledge synthesis is an art that depends on the free-wheeling individual minds and well-balanced views of the research mission.

How can we systematically nurture knowledge synthesis in future scientists and engineers? This is a fundamental question for educators not only in Japan, but in other countries as well. Science policy makers, industry leaders, and academics must understand that funding fundamental scientific research programs contributes to enriching only one component of the infrastructure for technological advancement. And scientific research conducted with general objectives, such as environment preservation and energy saving, rarely produces technological breakthroughs. Other components of the infrastructure need to be developed, but this is where innovations in policymaking are needed. Those infrastructure components - that is, the mechanisms to channel social needs to well-defined scientific programs and the initial funding sources - are time-dependent. It is impossible to reproduce, for example, the infrastructure that worked well in the U.S. in the 1950s. Japan must assess the present potential to create a new infrastructure.

Japan has benefited from the energy of grass-roots entrepreneurs, as described in Chapter 3. Most small manufacturers that flourished as venture businesses in the late 1940s and throughout the 1950s were gradually folded into the hierarchical organizations, called *Keiretsu*, with large, well-known companies at the top. In the late 1980s and the early 1990s the outsourcing of manufacturing components from parent companies to contractors or subsidiaries in Southeast Asia and China became widespread, eliminating those domestic small manufacturers near the bottom of the hierarchy. In the early 1990s, the bankruptcy rate of small manufacturers has exceed the rate of start-ups, signaling the danger of extinction of grass-roots entrepreneurship in Japan. The MITI is well aware of the need to rekindle the energy of small manufacturers by providing economic incentives [Fukuda, 1997]. If such a policy is integrated and coordinated with the growing fund for scientific research, there may emerge a unique mechanism for technological advancement in Japan.

The intensity of competition within the society has a strong relationship to the mode of technological development. Too much competition, whether between companies, government agencies, or individual scientists, forces competitors to avoid high-risk ventures and seek quick profits. This, in turn, encourages science and technology forces to take the course of incremental progress. On the other hand, the absence of competition, either in a monopolized industry or due to lack of outside interest, stalls progress. Japan has a unique combination of tough industry competition and little competition in the education system. The domestic electronics market is an arena of overly intense competition. One of the key issues for policy makers is how to create a reasonable level of competition in all sectors of the society. This does not mean reducing competition through government regulation; everything must be achieved through a free and open market. The government's role is to provide an infrastructure that widens the market where competition is too tight. For example, the infrastructure for information-age communications lags well behind those of other countries. Abundant and economical communications channels can foster the growth of the user population and diversification of products, thus diffusing the competition over a narrow range of products; most importantly, it can nurture inventors who will generate communications software and hardware of real impact.

5.3. Investments in the Electronics Industry

Over the last two decades, Japan has maintained approximately an 80-20 percent split between industry and government funding of R&D [Foreign Broadcast Information Service 1993]. In 1991, the percentages were: 80.6 percent industry, 11.5 percent universities, and 7.9 percent government.

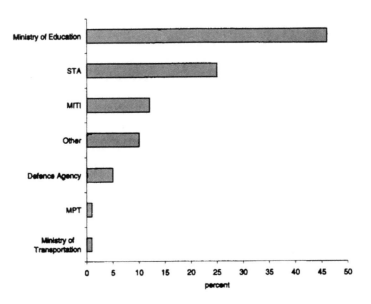

Figure 5.8 1993 Japanese R&D funding by agency (Total $2,226 billion).

However, since almost all Japanese universities' research funding is supported by the government, the last two numbers are often combined.

The Japanese government spends relatively little on R&D, compared with Japanese industry. For example, in 1991 the Japanese government spent approximately $20 billion compared to $828 billion spent by industry. However, government spending on R&D does have an influential and significant effect on the amount and type of R&D undertaken in Japan. This funding sends signals that certain technology development fields are important or are a high priority. Government influence also is achieved through policy directives: funding of cooperative R&D, subsidized and conditional loans, modest investments in high-risk projects with commercial payoffs, and tax incentives [Cheney and Grimes 1991].

According to the Agency of Industrial Science and Technology (AIST) of MITI, three agencies account for 83 percent of all government spending on R&D: the Ministry of Education (46 percent), the Science and Technology Agency (25 percent), and MITI (12 percent). Other relevant agencies funding R&D include the Defense Agency, the Ministry of Posts and Telecommunications (MPT), and the Ministry of Transport, as shown in Figure 5.8. MPT has relatively little budget authority but still funds telecommunications and related information technologies. The attention

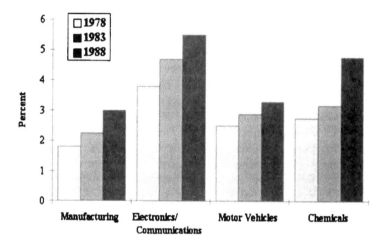

Figure 5.9 Trends in R&D as a Percent of Sales for Various Key Industries.
Source: Japan Management and Coordination Agency

given to electronics by Japanese funding programs is reflected in Figure 5.9, which depicts the trends in R&D as a percent of sales in various industries.

In terms of industrial funding, a MITI survey in 1992 of 1,667 companies found total plant and equipment investments in Japan valued at Yen 20.8 trillion. Of this, electronics equipment firms, including components firms, invested Yen 1,399 billion [Kelly et al. 1995]. Production-related investment accounted for 45 percent of overall outlays for new and high-value-added products. R&D-related investment represented 24 percent of overall outlays, with a priority on supplying competitive products. The most significant product-focused investments were in LCD production. Notebook-type personal computers posted significant increases in investment. Investment in 4M and 16M dynamic random access memory (DRAM) semiconductor capacity was significantly reduced in 1992. Investment in cordless and portable telephone capacity was heavy, as demand continued to grow. Computer-related R&D investment was also high.

5.3.1. R&D Efforts

To encourage R&D efforts in the electronics industry, Japan developed its own programs of joint government and industry funding late in the 1960s. The first successful project titled very large scale integration (VLSI) was aimed at making the Japanese semiconductor industry competitive in global markets at a time when it was behind the American industry. The research focused on digital integrated circuits and manufacturing processes and helped to raise Japanese microelectronics makers to the high-technological level. MITI and five top Japanese microelectronics companies (Hitachi, NEC, Fujitsu, Mitsubishi, and Toshiba) participated in this $300 million project over a five- year period. Half the budget was provided by the participant companies, while the net effect of government subsidies on the participants' overall R&D budget was incremental [Cheney and Grimes 1991]. Paralleling the VLSI project, Nippon Telephone and Telegraph (NTT) (when it was a government entity) conducted a VLSI project that focused on applications of microelectronics and telecommunications: "Wiring Technology: Multilayer Substrate With Photosensitive Polyimide Dielectric." In fact, this was a technology to make high-density multilayer substrates with a small propagation delay [NTT Technology Transfer Corp]. This effort was also successful, but there was not much coordination between the two projects because of bureaucratic politics. Another significant R&D effort in electronics was the Fifth Generation Project, started in 1981. Organized by MITI's cooperative research project, this 10-year effort bringing together the major computer companies in order to develop the "fifth generation" computers that followed vacuum tube computers, transistors, integrated circuits, and very large-scale integrated circuits. This project had at its core the Institute for New Generation Computer Technology (ICOT), with forty to fifty young corporate and MITI researchers who worked together on high-risk and high-payoff generic technologies. Young researchers (under age thirty-five) were selected because they tend to exchange information more freely and are characterized by more dynamism and more energy [Cheney and Grimes 1991]. ICOT was sponsored with at least a $40 million per year budget by MITI, because other companies were not willing to participate financially.

The work was conducted in parallel by participating companies at their own locations, involving about 100-150 researchers. The results were gathered as reports and meetings among researchers who also rotated in and out of ICOT. Even if the Fifth-generation computer-specific aims were too ambitious, successes like sequential-interference systems proved valuable [Cheney and Grimes 1991].

The "High-speed Computer Systems for Scientific and Technological Use" project, known as the Supercomputer project, was a nine-year program focused on research and development of GaAs and high electron-mobility transistors (HEMT). This project was sponsored by MITI and started with a total budget of ¥ 23 billion (approximately $115 million) in 1982. The specific goal for this project was to create a computer capable of performing ten billion floating point operations per second. Under a MITI contract, the Association for the Development of High Speed Scientific Computer Systems was formed by Fujitsu, Hitachi, NEC Corp., Toshiba, Mitsubishi

Electric, and Oki Electric Industry Co. All the work was performed in each company's facility and the information was exchanged through meetings arranged by the Association. For high-speed logic and memory devices, Josephson junction, HEMT, and GaAs devices were selected as possible alternatives to silicon devices [Park 1986]. Table 5-2 shows the budget allocated for the supercomputer project supplied by MITI. Packaging for these devices was also provided as part of the project. The project involved the following three key activities [Kashiwagi]:

- R&D of the parallel architecture and software
- R&D of new devices with high speed and a high level of integration
- construction and evaluation of the final system with the super-speed computing engine.

While there are other scattered examples of industry and government collaboration on specific electronics packaging R&D efforts, today two Japanese programs are crucial to facilitating this collaboration: Japan Key-TEC and Japan Research and Development Corporation (JRDC). They represent the fruits of cooperative government and industry efforts to fund R&D and have supported electronics (semiconductor) packaging and assembly research.

5.3.2. Key-TEC

In September 1985, the Japanese Diet announced the implementation of the Law for the Facilitation of Research in Fundamental Technologies. The Japan Key Technology Center (Japan Key-TEC) emanated as a direct result, received support from industrial and financial leaders, obtained the approval of MITI and the Minister of Posts and Telecommunications (MPT), and received initial capital investments totalling over ¥ 14 billion from various governmental and private organizations.

Table 5-2. The budget for the supercomputer project supplied by MITI (Yen 200=$1)

(JFY)	($ in mil)
1981	0.15
1982	4.1
1983	7.8
1984	11.2
1985	12.5
1986	14.5

The mission of Japan Key-TEC is to supply investments to meet the funding needs of various private-sector experimental and research projects, and to promote greater efficiency in such projects by coordinating joint efforts of researchers from government, industry, and academia. Japan Key-TEC's services include investments in joint R&D companies, conditional interest-free loans, and arrangements for joint research. In order to ensure that the interests of the private sector are considered, an Advisory Council of experts in various fundamental technologies is set up to act as consultative body.

Key-TEC has many well-defined ways of operating, depending on specific goals.

- Capital investment: Japan Key-TEC provides capital investment for fundamental research projects or comprehensive development projects in which two or more companies are jointly involved. In such cases, the companies must jointly capitalize and establish an R&D company. Japan Key-TEC provides up to 70 percent of the capital required for the R&D project.
- Loan services: Japan Key-TEC's conditional interest-free loan service mitigates the financial burden of unsuccessful research projects and reduces the risks and capital burden of research and development. An evaluation committee determines which projects to fund, based on the applications submitted. If the process doesn't prove to be significant, only the principal must be repaid. If the project shows valuable results, the project company will pay back the principal and interest equivalent to that charged for Trust Fund Bureau long term loans. The loan can be extended for up to 70 percent of the research project's costs, specified in advance. Any company (including foreign-affiliated companies) capable of performing such research is qualified to request this service.
- Mediation and execution of research and development: Japan Key-TEC offers mediation of services between private companies on the verge of conducting collective research with other national research companies, with a view toward promoting cooperation between government, industry, and academia.
- Japan trust international research cooperation service: Japan Trust Fund is a charitable trust that incorporates separate memorial trusts registered in the name of contributing individuals or corporations. Profits are used to invite foreign researchers in key technologies at a post-doctorate level to work.
- Research information service: Research literature and survey information on previously conducted research efforts are kept on file by Japan Key-TEC. Surveys among private companies and government

policy trends in various countries are also conducted by Japan Key-TEC in order to establish future strategies.

Since it was created, Japan Key-TEC has aimed to fund new and innovative approaches and technology in electronic packaging. For example, one of the first projects funded, Research and Development of Basic Technology for the Second-generation Optoelectronic Integrated Circuit, started in 1985 and extended until 1993. The project featured testing and research of crystals, processing techniques, and device technology for producing optical-electric integrated circuits (OEIC) with transmission speeds of 10 Gbit/sec.

5.3.3. Japanese Research and Development Corporation

The Japanese Research and Development Corporation (JRDC) is one of the key organizations implementing policies of the Science and Technology Agency (STA), which supports and "loosely controls" the corporation [Anderson 1984]. JRDC was established in 1961 to support advanced development in broad areas of technology. Today's JRDC is developing overall science and technology promotion systems in three main areas: basic research, technology transfer, and international research exchange. Funding for JRDC's activities is provided by the Japanese government.

Basic research: JRDC promotes basic research through Exploratory Research for Advanced Technology (ERATO), the International Joint Research Programs, and Precursory Research for Embryonic Science and Technology (PRESTO). Common to these programs is the emphasis on novel, open-ended research and the selection of young, talented individuals.

ERATO was begun in 1981 to give researchers the chance to enhance their ideas in flexible research conditions. JRDC selects key individuals with insight into a project and leadership skills as project directors. These directors put together a team of talented people from industry, academia, and government to work together in laboratories located to best serve the group.

It's mission was extended in 1989 for cooperative research projects between JRDC and foreign research organizations, which equally share costs and facilities. In 1991, JRDC began PRESTO to provide individual researchers with opportunities to conduct precursory creative research that could easily be carried out within conventional organizations.

Technology transfer: In 1986 JRDC began establishing High Technology Consortia in order to develop promising major research results that arose from ERATO, national laboratories, universities, and other sources. The objective is to transfer research results effectively, broaden the application of research results, actively promote development, raise the level of industry, and build better living. If a company can take the risk involved in developing a new technology itself, JRDC acts as a facilitator for technology transfer between the researcher and the company.

International research exchange: This program began in 1989 as one of several measures aimed at strengthening international cooperation in science and technology activities, in Japan. In this case JRDC implements the STA Fellowship Program, which provides foreign researchers with opportunities to conduct research at Japan's national laboratories and other organizations. To help researchers identify opportunities, JRDC provides information on Japan's research activities to foreign organizations and researchers. JRDC may also dispatch Japanese researchers from national laboratories to joint research programs with national laboratories in East European or Asian countries, providing researchers with travelling and boarding costs.

After research is completed, results that have potential for commercialization are developed further. If JRDC studies determine that a project is risky, it is directed to Cooperative Technology Development. The lower-risk technologies are directed to Technology Facilitation.

Technologies directed to Cooperative Technology Development are evaluated for novelty, economics, and public benefits. Those that cannot be easily developed by industry alone are given special attention. A Cooperative Technology Development contract is given to the companies best suited for the development. JRDC, the company, and the researcher that first provided the technology together plan the scale, time, and funding for development, and JRDC provides the necessary funds. The company develops the projects, receiving technical guidance from the researcher. JRDC supervises the activity to make sure the development proceeds smoothly and efficiently; at the end, it decides whether the project was successful according to criteria established at the beginning. If the project is successful, the company repays the funds over a five year period, with no interest charged; for a successful development, JRDC and the company reach an agreement for the developed technology, with the stipulation that JRDC will receive the royalties on sales of the resulting products. Half of the royalties are passed through to the researcher. JRDC also distributes the technology to other companies that have common interests.

Ideas directed to Technology Transfer Facilitation are publicized annually, described as available for licensing. JRDC also makes public presentations of new technologies with broad applications at about ten meetings held each year. JRDC's Technology Transfer Facilitation Commission transfers new technology through its members. If development results in saleable products, JRDC collects royalties, of which 8 to 90 percent is awarded to the researcher. If the results are requested by foreign companies, JRDC makes them available for transfer worldwide.

5.4. National Research Institutes and Cooperative Research

National Research Institutes (NRI) were created in order to support industrial technology development. In 1989, these institutes had a total budget of over $300 million [MITI 1989]. MITI's Agency for Industrial Science and Technology operates 16 of them. STA also runs some of these laboratories. Research within NRIs is part of a larger effort that occurs at many NRI locations; in these cases, much of the total funding for the project comes from the private sector. Coordination of these activities is provided by the sponsoring agency, in cooperation with leading firms [Cheney and Grimes 1991].

Cooperative R&D in Japan accounts for a low proportional of total spending. Sometimes, these groups involve 2 companies, sometimes large consortia. Sometimes MITI or another large agency brings companies together. Government policies, such as favorable tax and regulatory treatment or rapid depreciation of equipment are an encouragement for cooperative R&D.

5.5. University Research and Education

The pros and cons of the Japanese educational system have been discussed by a number of authors — for example, Lorriman and Kenjo [Lorriman and Kenjo 1996] and Cutts [Cutts 1997]. This section examines those aspects of education relevant to the electronics industry.

To understand the state of the art of Japanese universities, three factors need to be considered: faculty and staff, research funding, and graduates. These must be viewed against the backdrop of a hierarchical organization that has been maintained since the era of modernization in the late nineteenth century.

Their curricula, organization, faculty employment, and student enrollment programs are approved or reported to the Ministry of Education. The university with the longest history in Japan, the University of Tokyo, was founded in 1886. Subsequently, six universities were established in Kyoto, Tohoku, Kyushu, Hokkaido, Osaka, and Nagoya between 1897 and 1939. These seven universities were called imperial universities until the end of World War II; today, called "former" imperial universities, they still command high national respect. Added to this top group was the Tokyo Institute of Technology, which in the pre-war years attained a unique standing by specializing in textiles, ceramics, and electrical, mechanical, and civil engineering. In addition, Keio University and Waseda University, both private institutions, have histories comparable or longer than the imperial universities, and are prestigious. Keio was founded by Yukichi Fukuzawa, a pioneer of modern business in Japan and thus is well known for its history of producing successful businessmen. Waseda was founded by Shigenobu Okuma, a powerful politician in the Meiji era, and attracted students with political aspirations. Today, both Keio and Waseda have strong science and engineering departments.

Besides the eight national universities described above, there are 128 national universities created after World War II. Most of them are located at the prefecture capitals and other major cities. A large percentage of them are former vocational schools that were upgraded after WW II. Besides Keio and Waseda, there are 375 private universities, some founded in the pre-war years, but mostly recent arrivals.

In the 1990s the Ministry of Education began the process of upgrading the eight old national universities to the rank of "graduate study institution" (*daigakuin daigaku*), in order to direct the growing funding for fundamental research there. According to the plan, the enrollment of graduate students will increase, while undergraduate classes will be reduced. Consequently, the school system will have one more layer at the top of the hierarchy.

5.5.1. Faculty and Staff

The faculties of those top universities are mostly occupied with their own graduates. A professorship at these universities is a highly regarded position, so only those outstanding scholars who could enter these universities are entitled to such positions, so Japanese thinking goes. This means that elite professors entered the university in their late teens and never gained work or teaching experience elsewhere. They are competent individuals, but lack of experience in the outside world hampers their sensitivity to industrial needs. The elite of these elite professors serve on the science policy committees of the government and present their perspectives about the future of science and technology. They also serve as referees for research proposals submitted to the Japan Society for Promotion of Science (JSPS), the equivalent of the National Science Foundation in the U.S. The progress of fundamental science may not suffer as a result of their recommendations, because fundamental science can be pursued with little regard to industrial needs. What will suffer, however, are policies for technological development and research by engineering faculty. Oblivious to industrial needs, even engineering professors are bent on fundamental scientific research, and propagate and follow those recommendations to pursue it. This bias toward pure science pervades the engineering departments in other universities, and is propagated through the flow of professors in the nationwide university system and funding by the JSPS. For example, the mainstream research for electrical engineering faculty has been the solid-state physics related to electronic devices; there are few instances of research on such topics of industrial interest as routing and packaging.

In an attempt to enhance ties between the elite universities and industry, the Ministry of Education is encouraging the introduction of endowed chairs financed by companies. The company-backed endowed chair is allowed to function for three years, with a possible extension of a few more years. In many instances the chair is occupied by well-known scholars invited from the U.S. or Europe who are on sabbatical leave from their home schools. Interactions between the chair professor and the other members of the faculty are confined to a small sphere. Often, only those who worked to bring in the scholar collaborate with him or her. The search for candidates for a chair is conducted by those who took the initiative to establish it, so the invited professor is often an expert in the same field as the host. Unless the host is sensitive to industrial needs, collaboration between the host and the chair holder is confined to a narrow scholastic area. When the host is conscious of the need to break the old mold of the established university, a professional engineer from the industry may be invited to occupy the chair. In this case, the invited professor initiates industry-related research programs. However, when the time comes to fold up the chair, such programs often leave no trace of influence on the mainstream research.

When the university system was expanded in the early part of the century, a mechanism was devised to transfer advanced knowledge from the top universities to lower-level universities. Professors at top universities leave when they reach the mandatory retirement age, then go to teach at lower-level universities. The retirement age was set to facilitate the flow of retirees from one level to another in a stepwise fashion, and has been maintained to this day. The mandatory retirement age is 60 at the University of Tokyo and the Tokyo Institute of Technology, 63 at the other former imperial universities, 65 at most of those universities established after World War II, and 70 at many of the private universities. This mechanism also serves to vacate coveted professorial chairs at prestigious universities to younger generations.

This system served the nation well when it needed to catch up with the technology of the West. The elite professor brought knowledge of Western technology acquired through study abroad, reading publications, and follow-up experiment or analysis. The system produced some talented scholars who made genuine contributions to the advancement of science and technology, as testified by a few Nobel prizes. However, the majority of elite professors served as knowledge importers whose information seeped down through the hierarchy of the school system with the flow of professors.

Another feature of Japanese universities is rooted in the imperial universities: the *koza* system, in which a full professor, an associate professor, one or two research assistants and, in some cases, a technician, form an autonomous unit. The full professor is the master of knowledge, and he imparts his knowledge to his disciple, the associate professor. The associate professor is a senior brother to the research assistants. The research assistants are senior brothers to the students who come to the *koza* for thesis work. The technician used to serve as a drill sergeant who knew nooks and crannies of how to run the laboratory, but this species is close to extinction today. According to traditional Asian custom, the disciples do chores while learning from their seniors. The partnership lasts for ten, twenty, thirty years, depending on the age difference between partners. When the personal relationship between the master and the disciple becomes strained, either by differences in scientific opinion or personality, the master arranges transfer of the disciple to another, often lower-level, university. The Ministry of Education is now out to dismantle the *koza* system, which has been criticized for killing the originality of the young. The University of Tokyo and others adopted the big *koza* system, where associate professors and research assistants are granted more mobility in partnership with senior professors. However, personal partnerships that have been cultivated over the years do not disappear quickly, except in cases of personal discord.

The intimate personal ties across generations leave negative effects on engineering research in the current rapidly changing technological environment. The master may be an expert on the transistor, so his legacy almost forces the next generation to continue work on transistors. The entire system encourages incremental progress in extending existing technologies. And, since solid-state physics has been an essential discipline to understand transistor behavior, the interest of the next generation is likely to be riveted to solid-state physics. Research on the transistor is still an important topic, but other developments shape the entire organization of electronics technology. Many of those researchers from advanced countries who have scholarships or fellowships to stay in Japan want to find work experiences in companies, not in universities. They know where real work is going on.

People do indeed leave industry for the universities. The universities, in this case, are lower-rank national, prefecture, and private universities. The majority of those who leave the industry for teaching jobs are technical professionals who reached or came close to the mandatory retirement age (55 to 60) in the blue-chip companies. They know about real technology, and teach manufacturing, packaging, circuit design, and other industrial topics. Their research funding is scarce and the research facility is often inadequate; besides, they have to teach many classes and supervise over ten undergraduate students for their fourth-year thesis work. This is certainly not an environment in which any substantial breakthroughs can be expected.

5.5.2. Research Funding

Most of the funding for university research comes from the Ministry of Education directly or through the Japan Society for the Promotion of

Science. The direct funding for *koza* has been kept at the same level for thirty years, and the overhead charge for utilities, telephone, and other services is deducted from the *koza* fund. The *koza* fund is now close to nothing in regard to its function as a research fund. The JSPS fund has been increased in recent years, amounting to 100 trillion Yen in 1997. Proposals are reviewed once a year. Proposals focused on novel physical phenomena have a good chance of approval. The JSPS encourages applicants to organize a group study by university researchers to justify the allocation of large sums to research programs on priority subjects.

It has been pointed out that Japanese society has an egalitarian bent. When the *koza* fund was the sole source of research funding, disbursements were almost equal for all *kozas*. If the professor was not active in research, the fund was left unused and returned to the national treasury at the end of every fiscal year. More recently, there have been cries from active researchers for priority funding, which has grown to a substantial level. In 1995, the New Energy Development Organization, an organ of MITI, began granting priority funds, 100 million Yen per grant, on the basis of open competition. The fund by NEDO goes to such engineering research as new energy source development, energy conservation, and utilization of hydrogen energy. MITI and the Agency of Science and Technology are also expanding the budget for research grants. Priority subjects include superconducting materials, non-linear optical materials, quantum devices, micromachines, and bio-machines.

Japanese universities have long suffered from the trauma of the campus unrest of the late 1960s to the early 1970s. Collaboration with industry was regarded as negating the autonomy of academia. Funding of university research by industries almost stopped in the early 1970s. During the bubble economy of the 1980s the wall between the university and industry was breached through the donation of substantial funds for endowed chairs. In the 1990s industry funding is openly encouraged by the Ministry of Education. In 1995 the top eight universities amassed 28 trillion Yen for 6388 joint or contract research programs, and 21 trillion Yen in donations.

Although industry funding is increasing, funding per research program is modest compared to that on U.S. campuses. Japanese academics have complained about Japanese industries' willingness to donate huge sums to American universities. Japanese companies have supported research programs in high-tech areas by American university researchers, and some well-known companies donated chairs that cost up to ten times what a chair would cost in a Japanese university. The industries reasons are several. The most frequently heard is the close relevance of American research to industrial needs. Secondly, the industry wants to smooth relations with the prestigious American universities that exported knowledge and education to Japanese researchers in the past. On the American side, particularly in the

1980s, there were concerns about a possible drain of the fruits of scientific research to Japanese companies. Today, such concerns seem abated in the American economic boom.

5.5.3. Graduates

The rigidity of the Japanese university system is counterbalanced by the flexibility of its students and graduates. A typical engineering student spends the first year attending classes in language, art, science, and other non-engineering disciplines. Prevailing here is the popular theory among educators that engineers tend to be narrowly focused, and that teaching about humanities is essential to breed socially conscious engineers. This is a good time for most students to relax after intense preparation for university entrance exams, a process begun at age six or seven. The second and third years are relatively light. In the beginning of the fourth year, they are assigned to *kozas*, becoming members of the organization. The transformation from a free-wheeling youngster to a conforming, diligent organization member is swift.

About half of these students proceed to graduate school for two more years to earn a master's degree, continuing in the same *koza* or moving to another *koza*, depending on mutual competition and the quota for students per *koza*. The rest go to work in industry. Few continue to doctoral courses, except for those who aspire to a university career. Doctoral programs are dominated by students from other Asian countries.

What the company normally expects from the recruit is a particular mindset, not specific technical expertise. By some mental process experienced in the school years, a student acquires the mindset of an engineer. Although it is difficult to define "mindset," it implies a mental framework within which there is a lot of flexibility. In many cases the framework itself is so flexible that a mechanical engineering graduate can be transformed into an electronics engineer.

The real technical training starts after the graduate is employed. In most companies, the first two years of employment provide job acquaintance training. A supervisor who is a few years senior is assigned to each recruit. Training is scheduled, and a project is defined. At the end of the initial training period, trainees present the results of their assigned projects to an audience that includes top managers of the organization. This is also an occasion that teaches presentation skills.

After the training period, many other learning opportunities await. Most large companies have in-house schools offering various courses. Instructors are senior professionals of the company or professors invited from universities. On-the-job training is also an important component of the company education system. Besides OJT, evening classes are often held in a corner of the workplace.

With this system, it does not matter much what specific courses the engineer took at the university. This, in turn, means that statistics about graduates from engineering departments, and information about the curricula of these departments, have only slight relevance to the real education of the engineering population. This flexibility at the individual level is part of the reason why few layoffs of a generation of engineers are precipitated by the transition of technology from one generation to another. When, for example, COBOL is considered obsolete, programmers who have been working with that language start to learn C++.

5.6. A Vision of Miniaturization

SONY's continued success required more than product innovations. In addition to creating markets, SONY also maintained its market share vis-à-vis competitors like Matsushita, JVC, and Sanyo. SONY's strategy has been to develop and introduce next-generation products before its competitors. SONY defines next-generation products as half the price and one-third the weight or size of current products. SONY's past success has been based on new product developments based on this vision, such as SONY's portable compact disc players, the DiscMan and the MiniDisc. In 1998, SONY's current MiniDisc held a 40 percent share of the Japanese market.

The concept of miniaturization has spread in Japan, where lifestyles and small homes support rapid market acceptance of the concept. Small-sized TVs and stereos are suitable for typically small living spaces. The Walkman and MiniDisc have made it possible to listen to radio or tapes privately during the commute. SONY reduced the volume of its Walkman from 250 cc to 100 cc, approaching the size limit of a tape cassette. SONY's vision of portability and miniaturization has spread to most consumer electronics markets. NEC's cellular telephone was reduced in size from 600 cc in 1982 to 150 cc in 1991. Televisions using LCD monitors have shrunk from 650 cc to 400 cc. Notebook computers and hand-held "personal digital assistants" have eclipsed the early desktop computers introduced by Apple and IBM. Electronic notebooks have reached 50 cc in size.

Miniaturization trends have spread beyond consumer electronics into small-sized paperbacks that are easy for commuters to carry and read, thereby revitalizing a stagnant book industry in Japan. In the food business, package portions have gotten smaller as the average number of family members has declined. In clothing, down ski jackets have been reduced in thickness from 20 mm to 5 mm. In automobiles, Honda introduced its miniature "City" car. Even Japan's miniature bonsai trees got smaller with the introduction of "mini" bonsai trees. The *Nippon Keizai* newspaper described

Japan's miniaturization trend in 1982 with the term *kei-haku-tan-sho* (*kei*, lightweight; *haku*, thin; *tan*, short; and *sho*, small). SONY and other Japanese electronics firms continue to lead the miniaturization trend as it spreads around the world.

Since the introduction in 1985 of the portable 8 mm video camera/recorder, three successive generations of SONY 8 mm camcorders have weighed in at 1,200, 800, and 400 grams. Next-generation products require a broad range of technological developments. Camcorder developments provided the roadmap for Japan's electronic components and packaging industry for over five years. Every company in Japan uses a "roadmap" for future products that focuses R&D activities. As a TDK engineer noted, the SONY Handycam used to be the primary driver for miniaturization activities, then the cellular phone pushed development to 0.2 kg. The personal digital assistant (PDA) was the next multimedia product to push miniaturization.

5.6.1. Downsizing Electronics Products

The continuing miniaturization of portable electronics equipment requires supporting technological improvements. NTT's Mova (type TZ-804) cellular telephone, at 150 cubic centimeters, rivaled the Walkman or Electronic Calendar for miniaturization technologies. The reduction in size from 400 cc to 150 cc involved a reduction in all of the key components, including the antenna, receiver, controller, battery (through power management methods) and body vacancy (by using 0.4 mm thin parts, 0.5 mm pitch LSIs, and ultrasmall 1005 parts). Of the 250 cc size reduction, approximately 150 cc came from electronics packaging technologies and about 75 cc came from the integration of functions with semiconductor designs. The continued integration of functions within ASIC designs and the introduction of smaller 0804 parts has reduced the battery needs and body size another 75 ccs.

Miniaturization of parts and components is clearly an integral part of next-generation products. As shown in figure 5.10, Hitachi's goals for application-driven electronics packaging requirements are higher performance, speed, and power; smaller, thinner, and lighter packages; higher-density assembly and advanced automation; and multiple package types and integrated multichip module systems. SONY cut the number of parts in the optical pickup used in its MiniDisc from eight to two, cutting both its weight and cost drastically. Innovative designs and high levels of electronic integration can cut costs from 30 to 70 percent.

5.6.2. Japan's Roadmaps

Products like camcorders, palmcorders, VCRs, and cellular phones have been driving the development of electronic packaging and corresponding assembly technologies. These products utilize a relatively large amount of analog circuitry, which has required Japanese suppliers to develop cost-effective, high-volume processes for assembling passive components. The Handycam and cellular phone have provided the roadmaps for technology development. SONY, for example, set half-size targets for next-generation products which have been well communicated to suppliers, who now have future specifications for electronic packages and components. The use of 0804 packages and smaller component formats requires both leading-edge surface mount process capabilities and ultra-small component development.

Japanese vendors produce both components and assembly equipment and allow for high-density SMT assemblies using fine-pitch ICs, small passive components, and fine-pitch SMT connectors. As shown in Table 5-3, ceramic conductors and resistors are typically SMT components. Technology trends are shown in Figure 5.9 and 5.10. To meet future miniaturization needs, passive component suppliers, like TDK, have developed three-dimensional, multilayer, multicomponent ceramic modules that incorporate passives and allow discrete components to be assembled on the surface [Okamoto, 1994]. Cellular telephones and personal-sized computers are key applications. Rapidly growing LCD production may be

Table 5-3 Surface-Mount Applications

	Ceramic Conductors	Resistors
Headphone stereo	100%	100%
Cordless telephone	100%	100%
Video camera	95%	100%
LCD TV	85%	100%
Notebook PC	60%	50%
CD player, radio, cassette tape player	50%	40%
Desktop PC	50%	35%

the next product to drive ultra-small assembly technology. For example, NEC is developing a 0.1mm pitch assembly for applications in LCD driver production. The primary technologies used for electronics manufacturing are shown below [Boulton, 1995].

Multi-chip module (MCM) is growing slowly, with about 7 percent of Japan's current packaging applications. In computer applications, the rise in clock speed is pushing the shift from laminated PWBs to ceramic PWBs to SOS (silicon on silicon) as through-hole vias drop from 1.0 mm to 0.1 mm and smaller. At the same time, MCM-L applications will continue, as higher frequency demands push up the use of ceramic co-fired MCM (MCM-C). At the high end, as frequencies push above 100 MHz, it is expected that we will see a growth in the use of ceramic thin-film MCM (MCM-D/C) and MCM-SOS [NEC Corporation projections]. As we look at the rapid integration of functions into evolving electronic products, we can expect an increasing merger of technologies like integrated circuits, electronics packaging, and flat panel displays. The next generation of flat panel displays will have to cut costs and weight substantially, from the current $600 active matrix displays to $200. Such a reduction will require new technologies like plastic or glass-on-chip. In 1998, Sharp announced its plans to produce super twisted neumatic liquid-crystal displays (LCDs) using plastic instead of glass for lighter and harder-to-break panels. The new display weighs 25 percent of glass panels the same size and weight. Sharp intends to begin using them in cellular phones, digital assistants, and portable notebook computers. Sharp

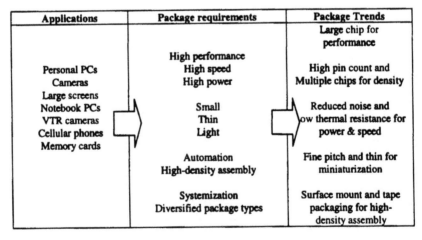

Figure 5.10 Market-driven Demands for Electronic Packaging
Source: Hitachi Corporation

Table 5-4 Electronic Technology Trends

Technology	Standard	Advanced Technology
Plastic/Ceramic Packages	Thin QFP TCP	Super-thin QFP/TCP Ball-grid Array Mullti-chip mounting/MCM
Discrete Components	1.0 mm x 0.5 mm	0.8 mm x 0.4 mm or multi-layer modules with built-in components
Printed Wiring Board	Laminated PWB 0.6mm thickness 127 micron lines 0.3mm vias 6-8 layers	Ceramic PWB or Silicon-on-silicon 0.2mm thickness 50 micron lines 0.1 mm vias 8 layers
Surface Mount	0.4 mm pitch	0.15 mm pitch

hopes to gain 20 percent by 2000 of the expected 210 million unit market for cellular phones and pagers [*The Nikkei Weekly*, 1998].

Japanese companies like Sharp and SONY have articulated their technology strategy to develop key components that can be applied to a wide range of products. SONY, for example, manufactures about 65 percent of the key components for its compact disk player and 45 percent of the components for its 8 mm camcorder. Advances in LCDs, CCDs (charge couple devices), and electronics packaging technologies were all needed for realization of camcorders and compact or minidisks. Using an LCD screen on Sharp's most recent 8 mm ViewCam helped it capture 25 percent market share in just one year. Competitive products have entered the market using similar technologies, typically including 1005 passive components, assembly densities of about sixteen components per square centimeter, and lightweight LCD modules. Continuous improvements in component technologies resulted in three new View Cam models being introduced by Sharp within one year of original EMS product introduction [Sharp Corporation Interview]. Mobile phones are now pushing the limits of low-cost packaging technology. SONY remains the leader in SMT assembly density, with twenty-five pieces per square centimeter for its mobile phones [Kaneda, 1994]. In 1998, Matsushita introduced its first digital camcorder using a mini-DV videotape. Weighing 440 grams, the NV-C was the industry's lightest vertical-designed digital camcorder. The new model featured 2 megabytes of flash memory, enabling it to take up to forty-eight still photos. A small battery pack allowed two hours of continuous operation. Victor Company of Japan also introduced a battery-powered portable color video printer weighing only 345 grams, which can instantly print from video cameras and digital still cameras.

5.6.3. Packaging Technologies for the 21st Century

In Japan, electronics companies are heavily invested in semiconductor technologies. Six vertically integrated firms produced 85 percent of Japan's semiconductors, 80 percent of its computers, 80 percent of its telecommunications equipment, and 60 percent of its consumer electronics products in the early 1990s. The primary technologies used for downsizing include continued miniaturization of surface mount devices, increased integration of circuit packaging, and reduced printed circuit board dimensions.

Surface-mount Devices. With the development of surface-mount technology (SMT), electronic parts such as transistors and capacitors no longer need leadwires. This reduces size requirements as well as potential quality problems related to bending, inserting wires into holes on PCBs, soldering, and cutting inserted wires. In the mid-1980s, 3216-type (3.2 mm by 1.6 mm parts) resistors and condensers began to replace the traditional leadwire parts for use in surface-mount applications. These new parts only require placement and soldering.

Surface-mount parts and application equipment have impacted the design and manufacture of most portable consumer electronics products, such as calculators, radio-cassette decks, cameras, and video cameras. The Japan Printed Circuits Association reported that calculators used 100 percent SMD for assembly. Products using over 90 percent SMD in assembly included radio-cassette decks, still cameras, and video cameras. The application of SMD to industrial products was growing, with larger computers using over 90 percent SMD, and computer peripherals, telephones, and electric parts approaching 90 percent.

The continued application of SMT allowed part sizes to be continuously downsized from 1608-type (1.6 mm by 0.8 mm) parts in 1990, to 1005-type (1.0 mm by 0.5 mm) in 1993, to 0804-type (0.8 mm by 0.4 mm) parts in 1996. This cut the initial length by half and the area by over 95 percent of earlier parts. As explained to the author by TDK:

"When we talk about things that are small in Japan, we call them like a grain of rice. Rice grains are 5.09 mm. Sesame seeds average 3.72 mm in diameter. In the early 1980s, the average size of capacitors and resistors were 3.2 mm by 1.6 mm. That was the same time that the next generation was introduced, but the 2.0 mm by 1.25 mm capacitors and resistors didn't become the standard until about 1990. Our 1608 multilayer ceramic capacitor of 1.6 mm by 0.8 mm was introduced in the late 1980s. We introduced the 1.25 mm by 0.6 mm versions around 1990, but don't expect it to become the standard until around the mid-1990s. The 1005, or 1.0 mm by 0.5 mm, was introduced in 1992 and 0804 in 1994. They are smaller than a poppy seed, which is 1.18 mm. People cannot handle such small size chips. We need machines to mount these small chips."

The weight of 1005 and 0804 is so light that the tension of solder exerts a force greater than the weight of the component. So they create new structural problems. Today, there are four machine makers that produce machines for 1005, but users are slow in buying them. These machines were introduced at the same time as the 1005 components through close coordination. The next generation is 0.8 by 0.4. Many people are addressing the solder problem and trying to develop reliable solder materials.

With vendors supplying both the components and the equipment, the time required for customers to utilize the new technology was reduced. Sharp, for example, designed new 1005 components into the PCB of its new 8 mm ViewCam which was introduced in November 1992. When asked if there were any problems introducing the new component technology, the plant manager said, "No, we just asked our suppliers to help us."

Integrated Circuit (IC) Packaging. New IC packaging technologies have increasingly been developed for SMT applications. For example, 100-pin quad flat packages with 0.65 mm lead pitch were introduced in the mid-1980s. The early 1990s saw 120-pin thin quad flat packages (TQFP) with 0.5 mm lead pitch and 1.4 mm thickness. Two-hundred-pin very small quad flat packages (VSQFP) with 0.3 mm lead pitch were introduced in 1994. Toshiba developed tape automated bonding (TAB) equipment for 520-pin, 0.2mm lead-pitch packages for sale in 1995. The new equipment utilized CCD image sensors to scan outer leads using a 400,000-pixel image density. Multichip module designs and ball-grid array packages also improved density for product miniaturization.

Oki Electric Industry Company, a major supplier of electronic components, was especially helpful in providing roadmaps for logic LSI packaging, memory LSI packaging, and tape carrier packages. The Oki roadmap for logic LSI packaging is shown Table 5.5.

Printed Circuit Boards. By 1991, single-sided rigid PCBs represented about 14.4 percent of the applications. One-third of the applications were double-sided rigid PCB applications. But to improve density, the use of multilayer PCBs is growing rapidly, with doubling application in four years. Rigid boards with three to four layers accounted for 20.3 percent of the applications in 1991, with five-to-nine-layer boards accounting for 14.4 percent, and ten-layer boards accounting for 3.1 percent.

With the increasing fluidity of product designs, there has also been growth in flexible printed circuits (PCs). One-sided flex circuits represented about 3 percent of applications in 1991, with multilayer circuits accounting for 5.6 percent of applications. New technologies for flexible PCs include flex-rigid and copper-clad laminate (CCL) for adhesiveless two-layer structures that enable high-temperature soldering at 350^0C for small electronic products.

Table 5-5 OKI's Logic LSI Package Roadmap (1993)

		1992	1994	1996	1998	2000
Devices Pattern width		0.5	0.4	0.3	0.25	0.2
Integrated density	DRAM	16M	64M	256M		
	SRAM	4M	16M	64M		
Size	Logic	250	400	600	700	
	DRAM	132	200	320	400	
Frequency	off chip	60	100	175	200	
	on chip	120	200	350	400	
Power	micro computer	3	4	4	4	
	mini computer	10	15	30	35	
	main frame	15	30	40	100	
Bonding Pitch	Wirebond	100	70	50	limit	
	TAB	100	70	50		
	flipchip	200	200	150	100	
Package smaller/thinner	thickness	1.4	1		bare chip	
	pin count	<80	<208		MCM-L	
	pin pitch	0.3-0.5	0.3-0.5			
	power	<1W	<2W			
High pin count	pin count	200-300	300-400	bare chip		
	pin pitch	0.5	0.4	MCM-L		
	power	<2-5W	5-10W			
Area array	pin count	200-300	500-600	750	2000	
	pad pitch	0.2-1.5	0.2-1.0	0.15-0.8	0.1-0.5	
	power	<2-5W	5-10W	10-15W	40W	
Application	Work station	1 chip, PGA MCM (4-8 chip)			8-16 chips	
	Portable	1 chip/ surface mount MCM (area array and thin types)			COB COG	

Source: Oki Corporation

Chip-on-flex-board (COF) has become a key downsizing technology, with IC chips wirebonded directly to the flexible PC. This is used in NTT's most recent "wristwatch-style" pager. Camera makers incorporate the most extensive use of flexboard technology into their miniaturization efforts.

Matsushita provides an interesting example of video camera miniaturization efforts (Table 5-6). One of the measures used to assess the efficiency of design in PCB layout is the number of components per square centimeter. Matsushita recently described the improvements in the design and layout of its video camera between 1990 and 1993; Table 5-6 shows the increasing density of components used on the camera and video boards. Matsushita's product development strategy to cut its video camera weight by 200 grams and size by 100 cubic centimeters included:

- integration of functions to reduce the number of components

- increases in density of components assembly (8.5 to 13 items per square centimeter)
- increased functional integration and component density to allow reduced PCB size

Product roadmaps stress the continued move to lighter, thinner, and smaller products, enabled by advanced packaging and manufacturing technologies. The chip packages are shifting from quad flat packages (QFP), to chip-on-board (COB) and ball-grid array (BGA) components.
A study completed in 1995 (Table 5-7) found Japanese packaging technologies to be most successful in volume applications driven by consumer products. Today, Japanese firms continue to provide the key components and equipment for such applications, but electronic manufacturing services have been able to take the leadership in volume production. Even major Japanese firms now outsource much of their requirements. Casio even completes many prototyping activities in the factories of its key suppliers. Such trends will continue to put pressure on Japan to innovate more rapidly and continue to push technologies for the development of integrated multimedia products and miniaturization.

Table 5-6 Matsushita Video Camera Board Designs

		Si (6/90)	S5 (5/91)	S9 (91)	TI (4/92)	CS1 (4/93)
PCB size (sq. cm)	Video bd.	113	103	108	(one board)	(one board)
	Camera bd.	98	87	62		
	Total size	211	190	170	122	75
Number of Components	Video bd.	1100	1050	1050		
	Camera bd.	700	650	400		
	Total #	1800	1700	1450	1433	930
Average # of Component per sq. cm.	Video bd.	9.7	10.2	9.7		
	Camera bd.	7.1	7.5	6.5		
	Overall	8.5	8.9	8.5	11.7	13.0
Highest density		11	12	12	14	16

Table 5-7 Packaging Technology Assessment
(U.S. vs. Japan)

	Technology	Volume Applications
Single Chip		
Plastic	Japan	Japan
Ceramic	U.S.*	Japan
Multichip		
Thin film	U.S.	U.S.
Ceramic	U.S.*	Japan
PWB	Japan	Japan
COB, COG	Japan	Japan
Chip Assembly		
Flip-chip	U.S.	Japan
TAB	Japan	Japan
Wire bond	Japan	Japan
Package Assembly		
Processes, Tools, Density	Japan	Japan
Passive Components	Japan	Japan
PWB	Japan	Japan
Flex	Japan	Japan
Connectors (Elastomeric, Anisotropic)	Japan	Japan
Package Design	U.S.	U.S.

* Indicates status of IBM only in high-performance ceramics for single- and multi-chip
applications; other U.S. companies are generally behind Japan in this area.
Source: JTEC Panel on Electronic Packaging and Manufacturing in Japan, 1995.

Chapter 6

EPILOGUE

This book has described the dynamics governing the Japanese electronics industry. Defining forces shape the course of the industry while colliding and merging with each other. One of these forces is the intense competition at both the personal and corporate levels. At the personal level, it takes the form of grass-roots energy generated by entrepreneurs in the post-war years and entrapreneurs now filling the corporate design rooms and manufacturing plants. Competition between companies is driving the whole industrial organization to produce incremental technological progress at a rapid pace. Another force is conformity, which developed in a vertically oriented hierarchical social organization and still permeates all sectors of Japanese society. These two forces have been amalgamated to create an extremely efficient machinery for absorbing technological developments made elsewhere, polishing them, and turning out high-quality commercial products.

The future of the Japanese electronics industry depends on how well Japan can loosen the grip of conformity and foster the liberation of grass-roots energy for more diverse ends than the manufacture of end products. Worldwide, the market for knowledge-based commodities is expanding quickly, concomitant with the establishment of an infrastructure for the knowledge trade. Ideas, concepts, technical expertise — all these abstract items can be sold wherever they are needed. Unlike the work in manufacturing plants, intellectual work can be done by people of all ages. The knowledge-based industry is where Japan must turn for the well-being of its aging population.

The transformation of society, particularly of people's sense of value, is the most challenging agenda. In a centralized conformist society, people evaluate issues in light of the credentials given to them by the central authority. They are more receptive to ideas from above than to those from below. Even when ideas arise from below, their originators tend to vanish in the anonymity. In such an atmosphere, knowledge is given no commercial value.

Cultural transformation cannot be accomplished by mere criticism, but only through cool analysis and adjustment of the operative social dynamics. A little stimulus to the existing system could generate permeating effects that would eventually transform the society. This viewpoint needs to be shared by people at all levels, from policymakers to individual engineers. The issue of how to work on the inertia of the social system does not pertain only to Japan. In every nation and the world at large, technology is advanced by competition between social inertia and revolutionary energy. To engineer the technology for the betterment of the world community, we need to understand more about what is at the helm of the technology development.

References

Aida H., *Denshi Rikkoku; Nihon no Jijoden (Building the Nation on Microelectronics)*, NHK Publishing Corporation, Tokyo, Vol.1 (1995), Vol.2 (1995), Vol.3 (1995), Vol.4 (1996), Vol.5 (1996), Vol.6 (1996), Vol.7 (1996).

Anderson, A. *Science and Technology in Japan*, Esse Longman Publishers, England, (1984).

Boulton E. R., and Pecht M., *Electronics Packaging Technology*, December 1, 1994.

Boulton W. R., *Building the Electronic Industry's Roadmap*, JTEC Panel Report on Electronic Manufacturing and Packaging in Japan, International Technology Research Institute, Loyola College, 28 February 1995.

Cheney, D.W. and Grimes, W.W. *Japanese Technology Policy: What's the secret* (February 1991), Council on Competitiveness, 1-26.

Cutts R. L., *An Empire of Schools*, M. E. Sharpe, Armonk, 1997.

Economic Indicators, Far Eastern Economic Review, December 17, 1998, 58 and 59.

Feigenbaum E.A., and McCorduck P., *The Fifth Generation: Artificial Intelligence and Japan's Computer Challenge to the World*, Addison-Wesley Publishing Company, Inc., Reading, Massachusetts, 1983.

Foreign Broadcast Information Service, *Trends in Principal Indicators on R&D Activities in Japan*, Joint Publications Research Service, Washington, (November 19, 1993), 19.

Fransman M., *Japan's Computer and Communications Industry*, Oxford University Press, London, 1995.

Fukuda H., "Science and Technology Policy of Japan," *Kuramae Kogyou Kaisi (Alumni Journal of Tokyo Institute of Technology)*, April 1997 (No. 922), 21-29.

Fukuda M., "Trade partners press reform agenda", The Nikkei Weekly, October 19, 1998, 2.

Gover, James E., and Gwyn, C. W., 1992, *Strengthening the US Microelectronics Industry by Consortia*, Albuquerque, NM: Sandia National Laboratories, 18.

Handy Facts on U.S.-Japan Economic Relations: 1998, Japan External Trade Organization (Tokyo, Japan) 16.

Happoya H., Takahashi K., Igarashi Y., and Nishimura K. "Latest Packaging Technology for Mini-Note Personal Computer", *Preprint for 11th Meeting of Circuit Packaging Society*, 1997.

Indicators of Science and Technology (Kagaku Gijutu Youran), Agency of Science and Technology, 1997.

Integrated Circuit International, May 1997, 8-9.

JETRO White Paper on International Trade, 1997: *Global trade in the era of information communications* (Japan External Trade Organization, 1997) 36.

JTEC Panel Report, *Electronic Manufacturing and Packaging in Japan*, Japanese Technology Education Center, Loyola College, MD, February 1995.

Kaneda Yoshiyuki, "Future perspective of new consumer products and automation technology in global manufacturing environment", (SONY Corporation) July 7, 1994.

Lee C. S., and Pecht M., *The Taiwan Electronics Industry*, CRC Press, Boca Raton, New York, 1997.

Lorriman J.,and Kenjo T., *Japan's Winning Margins*, Oxford University Press, London, 1996.

Mainichi Shinbun (Mainichi Daily), *Interview with Experts*, January 6, 1998, 15.

Morita A., *Made in Japan*, E. P. Dutton, Inc., New York, 1986.

Nakayama, W., "Japanese Supercomputers in Thermal Perspective, " in *High Performance Computing in Japan*, Ed. R. Mendez, John Wiley & Sons Ltd., 1992, 55-73.

Nikkei Electronics, No.663, June 3, 1996, 82-90.

Nikkei Electronics, No.675, November 4, 1996, from the whole issue.

Nikkei Electronics, No.702, November 3, 1997, 69-72.

Nikkei Electronics, No.704, December 1, 1997, from the whole issue.

Nikkei Microdevices, July 1996, p.151.

Okamoto A., Miniaturization and Integration of Passive Components by Multilayer Ceramic Technology, 1994, TDK Corporation, quoted in Boulton and Pecht, *Electronics Packaging Technology*.

Peters L., "Japanese Technology: To Gigabits and Beyond," *Semiconductor International*, November 1997, 111-115.

Record of Round Table Discussion, Birth of Libretto 50 Reflecting Users' Demands, *Public Relations Material from Toshiba Corporation*, 1997.

Schaller R. S., "Moore's Law: Past, Present, and Future," *IEEE Computer*, Vol.34, 1997, 53-59.

Sterling K., *Electronics Manufacturing Services Industry for 1995, 1996, Chicago, Ill.*, IPC Report., 1-55.

Takahashi Y., "Progress in the Electronic Components Industry in Japan after World War II", *Technological Competetiveness: Contemporary and Historical Perspectives on the Electrical, Electronics, and Computer Industries*, ed. William Aspray, IEEE Press, 1993, 37-52.

The Bottom Line, Asia week, June 5, 1998, 78.

White Paper of Trade and Commerce, The Ministry of International Trade and Industry, 1996, 345.

Index

For Product Safety Concerns and Information please contact our EU
representative GPSR@taylorandfrancis.com
Taylor & Francis Verlag GmbH, Kaufingerstraße 24, 80331 München, Germany

www.ingramcontent.com/pod-product-compliance
Ingram Content Group UK Ltd.
Pitfield, Milton Keynes, MK11 3LW, UK
UKHW021609240425
457818UK00018B/459